Fossils, Dinosaurs and Cave Men
A Biblical View

By Patrick Nurre

The Northwest Treasures Curriculum Project
Building Faith for a Lifetime of Faith

Fossils, Dinosaurs and Cave Men
A Biblical View

By Patrick Nurre

Fossils, Dinosaurs and Cave Men, A Biblical View, 3rd Edition
Published by Northwest Treasures
Bothell, Washington
425-488-6848
NorthwestRockAndFossil.com
northwestexpedition@msn.com

Fossils, Dinosaurs and Cave Men
A Biblical View
Contents

How to Use this Study

When it comes to studying earth history and geology, it is vitally important that we help our students to develop a sound perspective about how this all relates to the Bible and especially to the Great Flood of Noah's time. That is the thrust of the *Northwest Treasures Curriculum Project*, of which this study is one component. **Bedrock Geology**, which is the heart of this program, helps to set the stage for studying all the varying components of geology, and is a great pre-study for this one. *Fossils, Dinosaurs and Cave Men* can, however, stand as a study on its own. Completion of this course, including activities, quizzes and the comprehensive exam are equal to one semester of Earth Science (Paleontology) credit.

Suggestions for use:
1. Keep a notebook that has a separate section for vocabulary. Each section of the study has words that are vital to understand. From these words, create your own glossary. Some of the definitions will be included in the text and some will not.
2. In your notebook, you will also need to keep a record of the answers to various questions at the end of the sections.
3. Outline each section, listing main ideas, with a few points of fact under each one.
4. If it is possible, try to get out into the field to find examples of the fossils you are examining. If you live near a volcanic area, take time for a field trip to research it and understand how it fits into this study of fossils.
5. Take the quizzes after each section is completed.

A possible plan might look like this (for each lesson):
Day 1: Study the vocabulary words in the section; add them to the glossary in your notebook; read text.
Day 2-3: Read text again; Hi-light and/or outline the paragraphs for the important ideas; discuss the lesson; do activities.
Day 4-6: If necessary, finish activities. Review text. Take Quiz if indicated. There is a final Comprehensive Exam.

Of course, none of the sections are the exact same length, so some might take less time, some more. This scenario, however, would anticipate about 1 ½ - 2 weeks for each lesson.

Patrick Nurre
The Northwest Treasures Curriculum Project
Building Faith for a Lifetime of Faith

It is recommended that you have a fossil kit to accompany this study. You can purchase one from NorthwestRockAndFossil.com. Samples should include the following, or ones similar to these:

Vertebrate fossils – dinosaur bone, fossil fish, Spinosaurus bone, Megalodon tooth, mammoth bone or tusk, gastrolith, shark or fish vertebra, Mosasaur bone

Invertebrate fossils – clam, sea urchin, limestone with fossils, gastropod, ammonite, belemnite or baculite, brachiopod, crinoid head and blastoid, insect, sponge, coral, bryozoa, foraminifera, worm tube, crinoid stems

Plant fossils – permineralized wood, fern, wood, opalized wood

Other – cast/mold, stromatolite, mudstone

Preface to Fossils, Dinosaurs and Cave Men

The topic of the Evolution of living things is an extremely complex idea. Even though it is taught as fact in most schools and in public life, it is far from being shown to be the scientific explanation it is touted to be. The idea of Evolution itself is in a constant state of evolution. Hardly a day goes by that an article does not appear in some news story about a discovery, overturning previously held "scientific" beliefs. (Throughout this book, you will often find the word *scientific* in quotes, as in "scientific." When used this way I mean either a bias that has been labeled as scientific, or a combination of scientific facts combined with a naturalistic interpretation.) Some of these include:

- Dinosaurs were lizards; dinosaurs were reptiles; dinosaurs were mammalian
- A theropod gave rise to Archaeopteryx who was thought to be the transitional link between dinosaurs and birds; then paleontologists decided that Archaeopteryx was fully bird; ancestors to birds were found in younger rock than Archaeopteryx (even though Archaeopteryx was fully bird); Archaeopteryx is not considered a missing link (something that links together living things of different kinds), but the search continues for feathered theropod ancestors to birds.
- 14 species of Triceratops to two species of Triceratops
- Misnamed dinosaurs; reclassified dinosaurs
- Changes in ideas of evolution via acquired characteristics to evolution via mutations to evolution via huge genetic leaps

Modern scientists would simply tell us that that is the way science works: self-correcting and reformulating. If that is true, why should Evolution be considered a closed fact of science?

Because of the plasticity of modern Evolutionary thinking, fossils are constantly being reassigned positions and skeletons are constantly being corrected to reflect current so-called scientific thinking. As such, a few of the fossil examples that I have offered as examples of Evolutionary thinking, may in fact be changed in the future according to the prevailing Evolutionary ideas. Ever since I have been a child fascinated with dinosaurs, hallowed ideas once thought to be scientific discoveries have been abandoned. Because of this, I will attempt to stay in tune with the continuing development of Evolutionary thinking and fossil discoveries and will from time to time, as necessary, edit this textbook.

While Evolution, as a scientific belief, is constantly changing, the Scriptures remain unchanged. The truth as taught in Genesis and elsewhere continues to provide answers to questions and to correct false ideas.

Introduction

The Book of Genesis is being attacked today by the established scientific community. The word science comes from the Latin word, *scientia*, meaning *knowledge*. No one who claims to be logical and knowledgeable wants to disagree with science, right? But this "science" is not real science. It is actually a false religion; a false knowledge that we call *Gnosticism*. The word Gnosticism comes from the Greek word *gnosis*, which means *knowledge*. The serpent in the Book of Genesis claimed to have knowledge of what God actually said and what He was doing. But it went counter to what God had actually spoken to Adam. That event in the Garden of Eden in Genesis chapter 3 is what we call *The Fall*. And although it was the one-time event that ruined creation, its formula has been repeated over and over again in various ways in Earth history ever since then. The science of today is a mixture of true science and that which is falsely called science. It is actually a type of Gnosticism that is attacking the true faith and true knowledge in a very clever and deceitful way – it is actually personal interpretation cloaked in scientific language. And because it is cloaked as science, it does not appear to be a false religion. Even many of the great Christian leaders of modern times have been and are being led astray by this false religion. Slowly but surely this attack on the Book of Genesis is eroding the very foundations of faith in what God has spoken in His Scriptures.

In the Book of Daniel, chapter 8, Daniel has a vision from God of the last days on Earth. Initially it is explained as represented in the king of Greece and the kings of Media and Persia. But it is also a type of what is going to happen in the last days. The goat mentioned in this chapter is a prince who will oppose the Prince of princes in the last days. What is important to note in this chapter is what this false prince does to gain his power at that time. Read this chapter and note his rise to power:
1. He magnifies himself to be equal to the Commander of the host, the Prince of princes.
2. He removes the sacrifice to the true God and throws down God's sanctuary on Earth.
3. He rises to power through transgression; he rises through sin and disobedience to the word of God. That, by the way, is what defines transgression. It is a violation of the rule that God has given.

4. He flings truth to the ground.

5. He impresses his will on Earth and prospers.

6. His transgression causes horror and tramples the Jewish people.

7. He attempts to reign over the Earth (this is called the *final period of the indignation* in Daniel 8:19.

8. He is the one who makes a particular man powerful and mighty.

9. He corrupts to an extraordinary degree.

10. He destroys mighty men and believers.

11. He causes deceit to succeed through influence.

12. He corrupts many.

13. He opposes the Prince of princes.

The serpent's rise to power in the last days through an unknown leader will be brought about through extraordinary deception and intrigue. But thank God he will be broken not by man, but by God Himself.

What is crucial to see from this passage is the evil and deceit behind the serpent's rise to power. This is what the mission of the devil (the serpent) is since his fall as an angel of light to his final attempt to overthrow God. He blinds mankind through lies and deceit to turn their faith toward something else. This same pattern has been repeated many times throughout history. But the most significant for our Western Civilization took place during the period of time known as The Enlightenment.

This Enlightenment from the 1700s through the 1800s was characterized by turning from the church and the Scriptures for truth and meaning. It was a time when man exalted his own reason and ability to discern his origin and his values. Man usurped the place that Scripture had held for generations. From that point man would explain the history of the Earth without reference to God or the Bible. This became what we now know as modern geology. Isn't it interesting that the very thing that verifies God's divinity, His power and His nature - the Creation, has been turned into the playground for the serpent? The Enlightenment began as speculation about the mystery of God's presence in the affairs of man and moved to alternate explanations about the history of the Earth and ultimately, the history of man. This speculation has turned into a false religion passing as science.

Overview of Fossils, Dinosaurs and Cave Men

This textbook will focus on fossils – fossils of invertebrates, vertebrates and man. What do the fossils really tell us? We will explore the cause, nature and result of the Genesis Flood in relation to the fossils and how fossils and dinosaurs fit into the Book of Genesis. We will also explore the so-called proof for the evolution of man by examining the fossil evidence that has been claimed to be representative of man's evolutionary development from an ape-like creature to modern man. And we will explore the phenomenon of cave men and how they fit into the Genesis picture.

Lesson 1 – In the Beginning

The Two Contrasting Views of the Origin of Life

Words to know: *fossil archaeology (archaeological) paradigm interpretation framework The Enlightenment biological evolution geology uniformitarianism cosmology materialistic naturalism*

   Darwin viewed all living things as coming from a common ancestor – without the aid of a creator. Therefore all living things are related. His *tree of life* illustration (represented by the image above) was powerful! It seems to give the impression that the connecting branches are real and therefore all things *are* related. Darwin also thought that not all humans were the same. Even within mankind, there were evolutionary differences. European intellectuals of Darwin's time honestly thought that they were the highest order on the evolutionary tree!

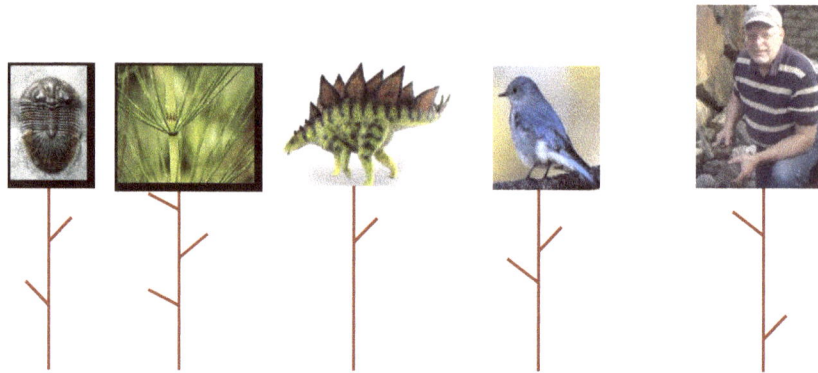

In the Biblical view of the origin of life, God created individual kinds of living things with the ability to produce variation *after their kind*. The fossil record bears this out. The hypothetical branches of relationship are missing. All known fossils show both variation and extinction, but not descent from a common ancestor. Man is a unique creation made in the image and likeness of God. Man has no relationship to the other animals.

What is a fossil? The dictionary definition states that the word *fossil* comes from the Latin, *fossilis*, meaning *obtained by digging* or *dug up*. A fossil is the preserved remains or traces of once-living organisms. The age of the remains really has nothing to do with whether something is a fossil or not. So, what are we digging up? Fossils represent life that has lived in the past, either a short time or a long time ago. There are a number of ways one can know about the past:

1) Written historical records
2) Eye witness accounts
3) *Archaeological* artifacts
4) Fossils

Fossils can give clues to the past, but they cannot tell a complete history of the past. In order to interpret the meaning of fossils, a prerecorded history of past life and events must be available or the *interpretation* will be subject to the ruling *paradigm* of the day.

The recorded history we will use in this study before us is the Bible. The Bible is accepted by almost everyone as a great moral book. But did you know that most of the Bible is a history book? The Bible records events and people that have taken place in the past history of the earth. These are not presented as nice stories or fairy tales, but records of what took place in the past history of the Earth. The very first verse of Genesis 1 starts with, "In the beginning…." It is meant to be taken as a historical record of what took place at the moment of creation. If Genesis was written to tell us great moral stories, but not necessarily anything literally true, it would have begun with, "Once upon a time." Fairy tales can convey lots of good moral meaning, but no one takes them as representing history. Genesis is different. Genesis portrays a history of the Earth, of sin, of judgment, of a global flood, of Israel and of a promised savior. Therefore, we can take Genesis as a historical **framework** through which to interpret fossils and artifacts that we *dig up*.

Now, it is very clear that the ruling paradigm of our day tells a different story. It does not use the Bible as a framework to interpret the data, but a philosophy that was sweeping the western world in the 1700s – **The Enlightenment**. The very words, *The Enlightenment*, convey the idea that a new way of thinking was introduced. The old way of thinking was the Bible. The new way of thinking and looking at things would from then on be through man's wisdom and his ability to figure things out. Old beliefs like a six-day creation, a devil, a global flood, an ark, miracles of healing and Jesus' own virgin birth and resurrection from the dead would be rejected in The Enlightenment. This way of looking at things eventually moved out of the realm of philosophy and into science in the forms of modern **geology** and **biological evolution**. Today instead of treating evolution as an alternate philosophy about the origin of the Earth and its living things, these new fields proclaim that it is the *only* true and scientific account of the origin of the Earth and life. Wow, how can you argue with science? This is why studying fossils is so crucial to our discussion.

Evolution as taught in science today means that the universe and all things in it, including life, morals, meaning, feelings and purpose have all developed over billions of years through natural processes – no miracles have been involved, only chemical processes. This idea was called **uniformitarianism** (the present is the key to the past). It is now referred to in many scientific circles as **materialistic naturalism**.

In the mid-1800s, the naturalist Charles Darwin, observed differences within certain kinds of animals such as different shapes and sizes of beaks within the finch birds. He proposed that given enough time, these changes would produce brand new living organisms. And he thought that the very young science of fossil study would eventually document this long journey of struggle and survival through the remains of once living things now preserved in rock. So, according to Darwin's own predictions, we should be able to find representatives of once living things that show that gradual, naturalistic change over time has produced the life we see today. The fossil record is rich with specimens – billions of them. And they continue to be discovered day after day. Certainly, we should at least stumble onto fossils of representative life forms that were in the process of evolving into other creatures! Our study in fossils will focus on past life and we will try to make some sense out of the evidence Darwin said would ultimately prove him right.

Charles Darwin as a young man, and a photo of the aged Charles Darwin. When Darwin took his famous voyage on the British ship, the *HMS Beagle*, he was given a copy of *Principles of Geology* by Charles Lyell. Lyell had been working on a naturalistic presentation of Earth history involving immense periods of time for the formation of the Earth. Darwin immediately saw the connection to his ideas about biological evolution –
with enough time, anything was possible.

Charles Lyell, British lawyer and geologist and his revolutionary book, *Principles of Geology*; it is still in print today. His personal goal was to remove any reference to Genesis from the study of geology or Earth history. Lyell believed that Earth history was not the result of some supernatural intervention of a god in the way of a global flood, but of long natural physical processes perhaps over hundreds of millions of years. Judging by the effect in modern geology and in the church today, I would say that he was successful in his attempts to do so!

New Geology

What Hutton and Lyell constructed was a new geology - a new way of looking at earth history. The scientists were no longer going to look at Earth history through the framework of scripture. The foundation for this new geology came out of The Enlightenment. It was an idea known as uniformitarianism formulated in the late 1700s. This big word meant that present physical geological processes should be and would be used to explain the past history of the Earth, not religious dogma such as the Bible. This new way of explaining Earth history would have far-reaching implications. Few realized this at the time, as it was being taught by respected men of society and means. And as science was the enlightened way of thinking, many of the great leaders of the church of the day began to reinterpret Genesis as a story with good moral meaning but not necessarily as a history of life and Earth. Many believed that science merely explained what was taught in Genesis. In reality this new geology was an alternate and contradictory history of the Earth compared to the one presented in the Bible. Coupled with Darwin's ideas, a sort of scientific Genesis was being written. Two completely different and competing views of life and Earth, also known as *cosmology*, the study of the well-ordered universe, were developed. This turned into a completely secular view of the universe. Today, this secular cosmology is being taught as scientific fact. The Biblical cosmology is being taught as myth and fairy tales.

Below are 2 sets of charts contrasting the views of Biblical Genesis and the so-called scientific alternative - *uniformitarianism*. The first chart compares the cosmology of the two views. The second chart compares the two views of the creation of life on Earth.

Comparison of Cosmology of Creation and Uniformitarianism

Genesis - Biblical Geology	Uniformitarianism - Secular Geology
Theism (God, but a specific God, the God of the Scriptures) - God has been actively involved in the origin and history of the Earth from the beginning; we call God's involvement in His creation *miracles*.	**Deism** (the deity; not identified, but sometimes referred to as *the Supreme Architect*) - the Deity may have created the earth and universe but if He did, He is no longer involved in it; He does not intervene. Only natural laws govern the history of the Earth; there are no such things as miracles
Gen. 1:1 in the beginning, **God**; *beginning* would mean the beginning of	**Big Bang** - in the beginning **matter** and space were already there; the Big Bang

time and of matter; God is eternal; He did not have a beginning. Another way to say this is, "In the beginning of our cosmos, God was already present. And it is He who has given order and meaning to creation."	does not attempt to answer the question, "Where did matter and space come from?" It is ignored.
Gen. 1:1 – God created the **space** and the **earth** on Day 1 of the creation (created means, *out of nothing*); Earth was the only *heavenly body* in the beginning.	Matter exploded, **forming** galaxies, stars, and planets 15 billion years ago; Earth came along about 4.6 billion years ago or 11 billion years after the universe came into existence.
Gen. 1:2 – the earth was originally a surging mass of **water** (the *deep*).	The earth was originally a cloud of **gas** or a **molten blob of magma/lava**; water came last
Gen. 1:2 – the **Spirit of God** was an active part of creation from the beginning.	God, spirit and religion are not matter and consequently are not relevant and are excluded from the study of Earth history
Gen. 1:3 – **light** (energy; the physical laws that govern the universe)	Physical laws and energy **evolved** as the universe evolved
Gen. 1:5 – *good* = perfect, done, complete in God's eyes; evening and morning define a complete revolution of the earth, a day; *one day* means there was one complete 24 hour period of Earth history with evening and morning (time divisions) and it was the first day of more to come	*Day* in the Scriptures is simply **symbolical** for a period of time; because of the undeveloped minds of these early people, cosmology had to be communicated simply; they could not comprehend *deep time*; according to evolution, nothing is ever completed, but always changing into the next thing over millions of years (*deep time*)

Order of Creation in Genesis vs. Order of Evolutionary Thought

Order of Creation in Genesis	Order of Evolutionary Thought
1:1 Time began when God created	Time is viewed as cyclical; no beginning and no ending; the universe in some form was always there
1:1 God Himself created; everything was brought into being from nothing by the word of God	The subject of God is irrelevant; everything unfolded naturalistically from other substances
1:1 The expanse or space (not the stars and heavenly bodies) and the Earth were created first	Matter and energy were present already and then other heavenly bodies slowly evolved beginning about 15 billion years ago and then the earth about 9 billion years after what geologists call the Big Bang
1:2 The Earth was developed first, but it was formless and empty	The Earth has taken about 4.6 billion years to develop, beginning with a molten ball at the beginning; all the necessary ingredients for formation and life were in the mix at the beginning
1:2 Darkness; there was no sun until Day 4 of Creation week	The sun (light) came first, before the Earth
1:2 Water was present at the beginning of the Earth's creation	Water came several billion years after the earth cooled
1:3-4 Light came into being by way of a direct word from God; God separated light and darkness; God was directly involved so our creation was a supernatural act	Darkness and light were there from the beginning of the Big Bang and is determined by celestial bodies like our sun; scientists decided in the 1800s that God was only involved in Earth's creation (Deism) so our Earth's development was a natural process occurring over billions of years
1:5 Light and darkness for the Earth was determined by 12 hour periods (evening and morning) and therefore one complete rotation of the Earth;	Light and darkness are a product of celestial bodies like our sun

verses 1-5 were accomplished in 1 literal day	
1:6-8 The Earth's oxygenated atmosphere was created on Day 2; water has oxygen and so oxygen was there from the beginning	The Earth's oxygenated atmosphere was developed over millions of years; in order for evolution to have occurred, the early atmosphere had to have been a reducing atmosphere (no oxygen) and so the waters were not oxygenated
1:9-13 The seas were gathered into one place and then the dry land appeared. The grasses, herbs and fruit trees sprouted with seeds on the Earth. The land vegetation was created on Day 3.	In evolutionary thinking the Earth is the focus. Plate tectonics of the land is what separated water into the oceans. Life began in the oceans. Grasses did not come on the scene for 5 ½ billion years.
1:14-19 The stars, sun and moon were created on Day 4 in relation to the Earth. The purpose for these lights from the beginning was to separate day from night, mark the days, seasons and years. God decreed this and set this up.	In evolutionary thinking the sun and stars came first, then the Earth and then the moon. Seasons, days and years, were a natural part of the expanding universe. God had nothing to do with it.
1:20-23 Life in the oceans and birds in the skies were created on Day 5, after plants were created on Day 4. Life's diversity in the oceans and in the skies was there from the beginning of their appearance. Life was divided into kinds. This is why we are able to classify living things. Kinds have diversity within them.	Life began in the oceans, then amphibians and reptiles and birds evolved from dinosaurs, long after life first appeared in the oceans. Many land plants evolved after land animals and birds. Life is viewed as a continuous phylogenetic tree with everything interrelated and interconnected.
1:24-30 Land animals, including reptiles, amphibians and mammals were created on Day 6, after the plants on Day 4, ocean life and birds on Day 5. Man was created on Day 6. He existed at the same time as all the kinds of land-dwelling animals which	The order of evolution: life in the oceans, amphibians, reptiles, birds, mammals and then man. Man came after the dinosaurs had died out. Many land creatures had existed for millions of years before man evolved. Man developed the idea of God and his

would have included dinosaurs. Man was created in the image of God as male and female from the beginning and is therefore unique. God's authority over man was placed there at the beginning of his creation. Man's authority over creation was granted by God and so man is a steward of God's things.	beginning through evolutionary development. He is not to maintain uniqueness among the other animals, as he is one of them. Females evolved as a separate line (although scientists dare not say which came first!). The idea that God evolved came after man was on Earth for a while. Man was not given authority over the Earth; instead, he usurps it. Everything is a connected system - nothing higher or more privileged than another.
1:31 The creation was finished by the end of Day 6. Everything was in perfect harmony with no death or struggle.	The *creation* is never finished. It is continuously evolving – a cyclical process. Death, decay and struggle are the vital processes of evolution, not of sin.

Now since we are dealing with past events that cannot be tested by science, we must represent the two contrasting views as frameworks through which to interpret the fossil evidence. And here is the question that we must ask as we proceed. Does the evidence from the fossils tell of a history of evolution from a common ancestor through slow and gradual transition from simple to complex life or does it tell of a history initiated by God with created kinds of living things with no gradual transition between the kinds? We will begin to explore the answer to this question in the next lesson.

You may be wondering how dating the fossils fits into the picture. In the early 1800s there was no scientific way to date the fossils and the rocks. This attempt would not come until about 60 years later. We will tackle that subject in a later lesson. For now, it is crucial that you understand the development of the thinking behind this new geology.

Activity 1

Obtain a college catalogue of science classes and their description. From this list, choose each class that in some way interacts with a uniformitarian viewpoint. The object of this activity is to become sharp at discerning within the sciences what actual science is and what is philosophy.

Please take Quiz #1, Appendix B

Lesson 2 – Life on the Early Earth

The Biblical Classification of Living Things

Words to know: flagellum   prokaryotic eukaryotic paleontology organelle Theistic Evolution   genetic Cambrian Explosion   paradox Paleozoic Era invertebrate vertebrate phyla polychaetes parapodia plastids membrane polyp sophistication hyolith laminae anthozoa secrete taxa cnidari protozoa cilia bryozoa primordial fauna

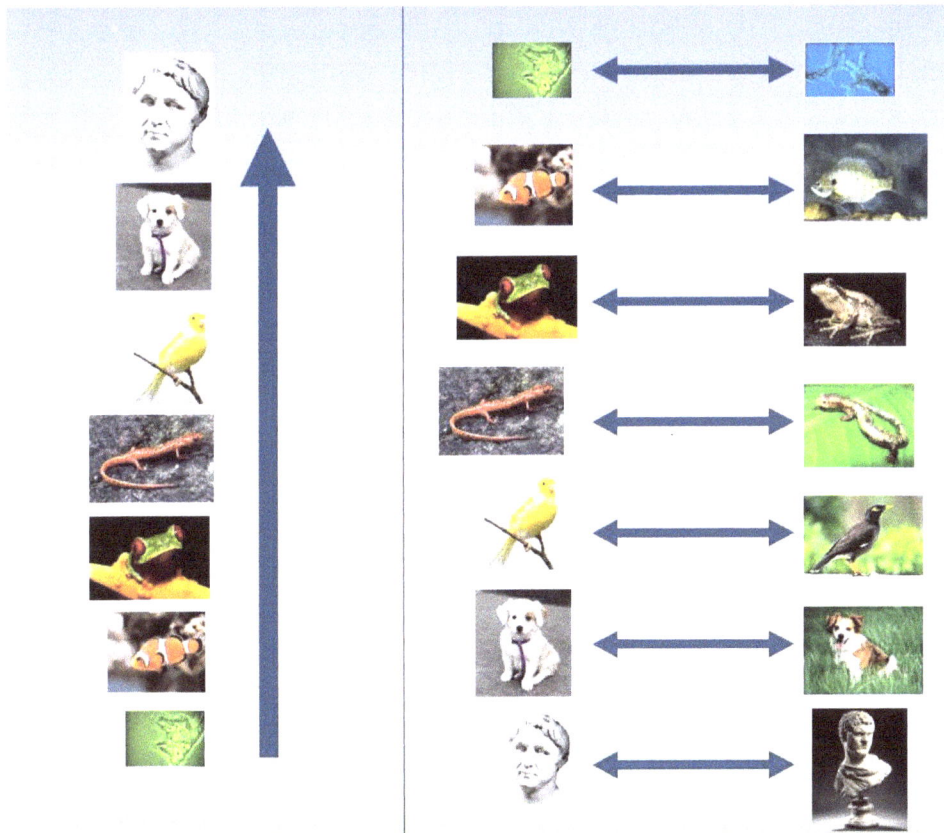

The evolutionary view of life   *The Genesis view of life*

For life to have evolved from simple life forms like protozoa to complex thinking man escapes not only comprehension but also the science. It is virtually impossible. The early attempt to classify living things teaches us that life was created as kinds with great ability to have a lot of variation within the kinds.

Technology and our knowledge of genetics have grown by leaps and bounds since Darwin's idea of evolution was introduced to the public in 1859. We know things today about living organisms that Darwin never even dreamed of.

One of the most critical facts of life is the amazing complexity and design that all of life possesses. Darwin's idea of simple life developing into complex living things is naïve and out of date. What we have discovered today, is that absolutely no life can be considered simple! Take the example of a living cell. Let's just take a small part of that cell called the *flagellum*. The dictionary defines the flagellum as a lash-like appendage that protrudes from the cell body of certain *prokaryotic* and *eukaryotic* cells. The word flagellum in Latin means *whip*. The primary role of the flagellum is locomotion, but it also often has the function as a sensory *organelle*, being sensitive to chemicals and temperatures outside the cell. Usually it is pictured like the one below on the left. But this does not show you how complex this little organ is or what must have had to take place in order for it to function with other coordinated parts. The picture on the right shows us how complex this little biological part is. The truly amazing thing is that it would not work unless all its parts were working at the same time and all the parts of the larger organism it is attached to are also working together at the same time.

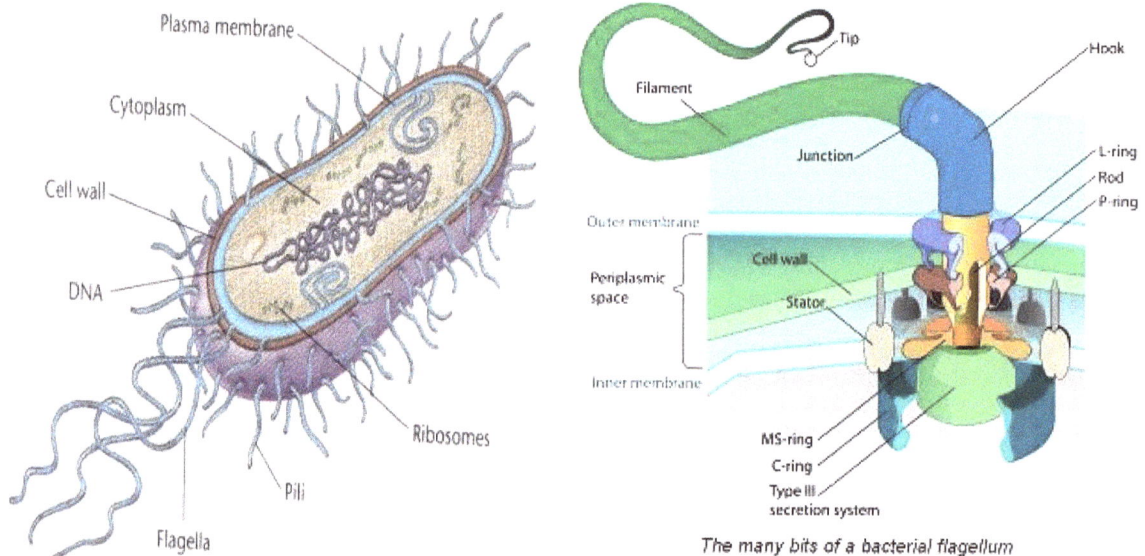

The many bits of a bacterial flagellum

The flagellum is actually a very complicated motor that serves a complex living system. And here is the absolute requirement – it must work together with all its various parts or it won't work at all. An organism with a design such as this demands an intelligent designer. How could something like this evolve all its movable parts over millions of years AND coordinate them so that they work together?

The only logical and reasonable explanation is that, "In the beginning God created…." An intelligent all-wise and all-powerful Being created life in all its complexity and design to work the way it does from the very beginning.

The Bible tells us that in the beginning God created kinds of plants and animals to reproduce after their kinds. Some have suggested that what God did was to create the first life and then let it evolve over hundreds of millions of years. This idea is known as **theistic evolution**. It was an attempt in the early 20th century to harmonize Genesis with the fledging Theory of Evolution. But this simply will not work. The same problems still exist. What is the specific process that developed one type of living organism into an entirely different one? The Theory of Evolution or as it is sometimes called, *Microbes-to-Man Evolution*, requires a gain in totally new information at each step of the evolutionary process. The **genetic** information for arms is different than the genetic information for eyes. To develop brand new systems and creatures there must be a source of new information and the systems already in place to utilize the new information.

The idea of evolution has never been shown experimentally to happen. Breeding is often brought up as proof for evolution. But what breeding demonstrates is that kinds of animals change within their kinds but not across kinds, no matter how much time is allowed. The other thing that breeding demonstrates is that this process takes away information. The different breed represents a loss of genetic information, not a gain of new directional information.

Here is an example. Suppose you want to build an engine for an automobile. You have most of the parts, except the pistons. Without the pistons of course, you will not be able to utilize the spark created in the piston chamber to move the rest of the parts needed to drive the car. Thus, you must have a system of parts that is working at the same time or the car will not run. The pistons will not form by themselves. Someone who knows (what we call intelligence) must design the proper part to work with the other parts available. He must also place the part in the right position in order to utilize the rest of the parts. The real question in uniformitarian evolution is how the various parts needed for life were developed and fit to work together with other parts to create or generate life.

The Cambrian Curtain
Does the record of the fossils we have collected over the past 250 years help us solve this dilemma? One of the greatest unsolved mysteries in **Paleontology** is the

event known as the **Cambrian Explosion** or **Cambrian Curtain**. Evolutionary biologist Jeffrey Levinton, though convinced of the common ancestry of animals, acknowledged in 1992 that the Cambrian Explosion - "life's big bang," as he called it - remains "evolutionary biology's deepest **paradox**." [1]

The Cambrian is designated by geologists as the lowest layer in the Paleozoic Era. The Paleozoic Era supposedly comprised around 295 million years of evolutionary history. The Cambrian Period, the lowest layer in the Paleozoic Era, has been characterized by geologists as a time with abundant fossils of multicellular organisms with shells which seem to have sprung up and diversified rapidly. Evolutionists believe that multicellular complex life must have taken hundreds of millions of years to evolve from their single-celled non-skeletal ancestors. To this day, however, the first skeletal and confirmed multicellular life has not been found below the Cambrian layers. The fossils in the Cambrian layers show up as completed, complex animals with variation right from the beginning. The few odds and ends that have been claimed to be fossil ancestors in the rocks below the Cambrian Period are unidentifiable "somethings." There are many opinions as to what these are, but nothing definitive has been concluded.

The *Cambrian* is a name for a particular layer of rocks in Britain that was studied by geologists in the early 1800s. It is supposed to have represented the period of time in Earth history when multicellular life with hard shells existed in abundance and variety. For many years, geologists have regarded this layer to be a major mystery. Life appears to have sprung onto the scene of Earth history fully developed and functioning. When geologists have looked at the rocks below the Cambrian, they have not found anything that even remotely resembles the Cambrian life. This is why this Cambrian layer is a paradox. Complex life with hard shells would certainly have required a long evolutionary history of development. But there is no such evidence of this history that has been found in the layers below the Cambrian. It has been called one of the greatest unsolved mysteries in modern geology.

---

[1] Levinton, J. "The Big Bang of Animal Evolution." *Scientific American* 1 Nov. 1992: 52-59. Print.

The Paleozoic Era

Age in Millions of Years	Period	Major Evolutionary Events
245	Permian	
290	Pennsylvanian	First winged insects
325	Mississippian	First reptiles First land vertebrates
360	Devonian	First seed plants First insects
410	Silurian	First jawed fish
440	Ordivician	First land plants
540-500	Cambrian	First vertebrates First multicellular creatures with hard shells – The "Cambrian Explosion"

Above is a diagram that is supposed to represent the evolutionary history of early life on Earth. This arrangement is part of what has been called the Geologic Time Table or Column. The section above is referred to as the Paleozoic Era. Paleozoic is from the Greek language and it means *ancient life*. The word *era* stands for a very long period of time, 295 million years of evolutionary history. The Paleozoic Era is made up of seven shorter periods of time called Periods – The fossil record in reality is more like Creation Week in the Bible in which all the various kinds of living things were created to reproduce with a great amount of variety within each of its kinds – but over six literal days, of course.

The lowest period in Paleozoic Era is called the Cambrian Period, so called for the rocks that were first studied in Cambria, Britain, the Latin word for Wales. The beginning of this period supposedly started 540 million years ago. It is at this point in time that most of the modern-day **phyla** of life suddenly existed and possessed variation with apparently no evolutionary history. It is as if a curtain of life was suddenly raised with fully developed, teeming life. The great mystery is that the amount of evolution that would have been required to produce this amazing array of life would have taken hundreds of millions of years to evolve, according to Darwin's ideas, but that evolutionary history is nowhere in sight. There are absolutely no undisputed fossils that would even begin to show how life developed from single cells to the diversity of multicellular, shelled life that abounds in the Cambrian Period. One evolutionary article explains it this way:

> At its (the Cambrian Period) beginning, multicelled animals underwent a dramatic *explosion* in diversity, and almost all living animal phyla appeared within a few millions of years.[2]

[2] Mark McGinley, University of California Department of Paleontology, topic editor, December 9, 2009, http://www.ucmp.berkeley.edu/paleozoic/paleozoic.php.

So what kind of creatures supposedly sprang into life in the Cambrian Period? Here is what the fossils show us:

1. **Brachiopods** - meaning *arm footed*, were bivalve marine creatures, representatives of which are living today, that resemble clams, but are entirely different. A brachiopod can always be identified by its shell arrangement. One half of its shell overlaps the other half, giving it an *overbite* appearance. Their evolutionary history is totally absent in the fossil record. Most of these seem to have gone extinct in the Genesis Flood or shortly after.

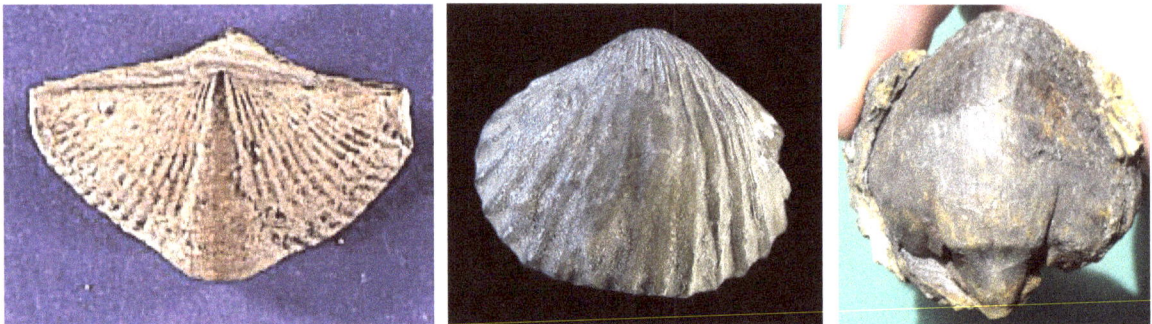

These brachiopods show an amazing amount of variation, but no evolution

2. **Echinoderms** – meaning *hedgehog skin* in the Greek language. The adults are recognizable by their (five-point) radial symmetry, and include such well-known animals as starfish, brittle stars, sea urchins, sand dollars, and sea cucumbers. All are represented today in living echinoderms showing absolutely no evolutionary change or ancestry.

The echinoderms show an amazing amount of variation, but no evolution

3. Jawless _vertebrate_ fishes – called _agnatha_ in the Greek, meaning _no jaws_.

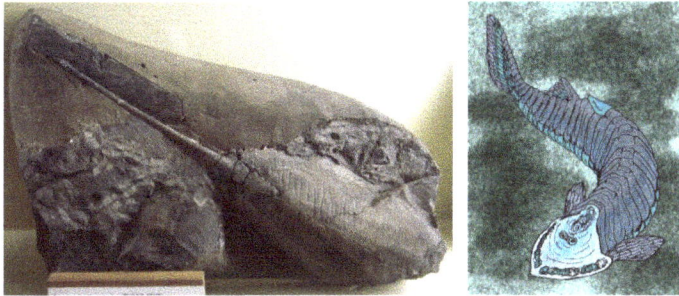

Geologists have identified this fossil as coming from the first jawless vertebrate fish, which is artistically rendered in the picture to the right. You may have trouble seeing this, but the important point here is that secular geologists have identified it as the _first fish_ from the Cambrian. This is very significant because it means that even vertebrate creatures show up in the earliest fossil-bearing rocks. This does sound more like the creation account in Genesis, doesn't it?

4. Trilobites – meaning _three lobes;_ Trilobites are the most prolific fossil of the Cambrian Period. It is now known that the trilobite had complex eyes and well-developed parts. Surely its evolutionary history must go back hundreds of millions of years. But no ancestor to the trilobite has ever been found below the Cambrian, and yet it was the most numerous of the Cambrian animals. It is generally accepted that one of the first and most complex animals to appear is the trilobite, an arthropod (defined as having joint-footed appendages), which by any reckoning must be viewed as a complex and morphologically advanced creature. This very complexity and sudden appearance suggests that the true origin was earlier than the fossil record indicates. As of this date, no evolutionary history has been discovered in rocks below rocks that have been identified as Cambrian.

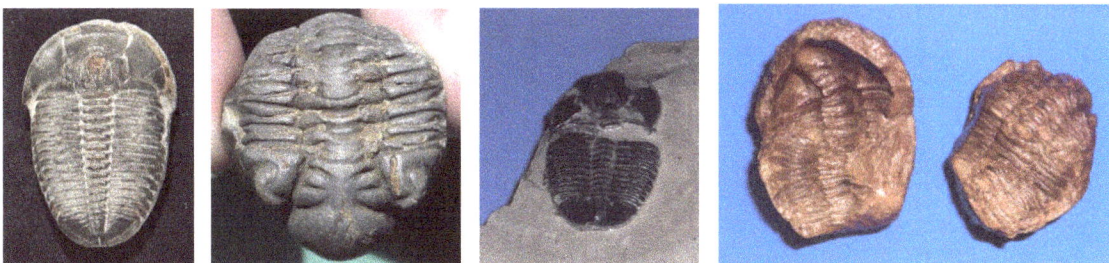

The trilobites show an amazing amount of variation, but no evolution

5. Mollusks – includes the cephalopods, bivalves and gastropods. Scientists tell us that the cephalopods were the most advanced of all the marine invertebrates. I am not sure exactly what that means, but certainly creatures of this description would have required hundreds of millions of years of evolutionary development. But here again there is no ancestry; not even a hint of evolutionary ancestry below the Cambrian Period. All this speaks of creation by an intelligent designer.

The mollusks show an amazing amount of variation, but no evolution

6. Annelids – (from the Latin, *anellus*, meaning *little ring*) are identified by geologists as the first segmented worms. These fossils are amazingly like the modern annelids. The basic annelid form consists of multiple segments. Each of these segments has the same sets of organs. These worms are called **polychaetes** – *meaning many bristles*. This means that each body segment has a pair of fleshy protrusions called *parapodia* that bear many bristles, called chaetae, which are made of a substance called chitin. Chitin is a very specialized cellulose material. It gives strength to things like the wings of insects. Polychaetes are sometimes referred to as bristle worms. Many species of these worms use **Parapodia** for locomotion. Polychaetes are essentially marine worms possessing both sexes and having paired appendages (parapodia) bearing bristles. In other words, this creature was already highly developed in the Cambrian Period – 550 million years ago, according to evolutionary theory. They must have taken millions of years to evolve. Yet, no ancestry has been discovered below the Cambrian Period rocks.

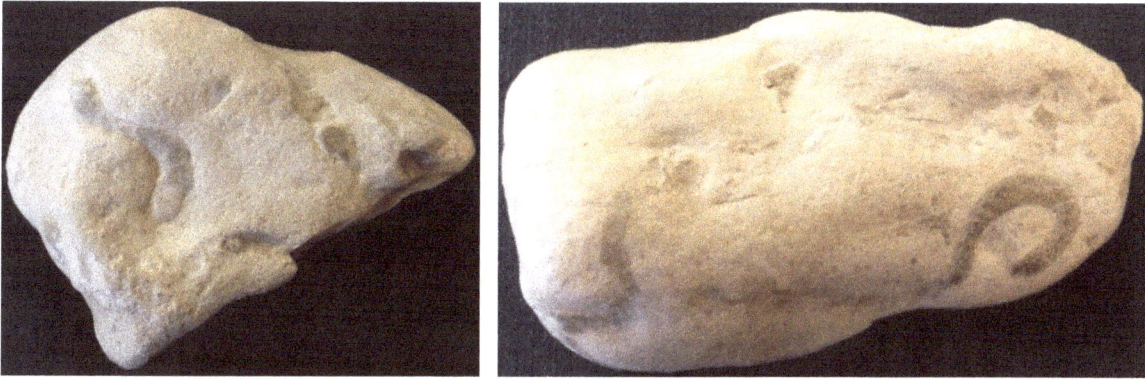

Worms seem to have risen suddenly and without any ancestors. They appear abruptly and in great variety from the very beginning of the Cambrian Period!

7. Red and Green Algae – algae derives its name from the Latin, *alga*, meaning *seaweed*. All true algal cells have a nucleus enclosed within a membrane. They also possess **plastids** bound in one or more **membranes**. Plastids are storage centers for cell products. They are also manufacturing centers that produce chemicals such as carbohydrates needed by the cell. This kind of complexity indicates a living organism with a high degree of **sophistication**. Certainly, these creatures didn't just evolve overnight! But again, there are no representative undisputed fossil algal ancestors to the algae, present in the rock layers below the Cambrian Period.

The algae show an amazing amount of variation and seem to have risen suddenly, but with no evolution.

8. Hyolithids – Now considered to be extinct, the **hyoliths** were calcareous (made of calcite), with an organic component, and had an organic-rich central core surrounded by concentric **laminae** of calcite; a coiled shell. The evolutionary development of the hard shells that mark many of the Cambrian Period is shrouded in mystery. No ancestry for this development occurs below the Cambrian Period.

29

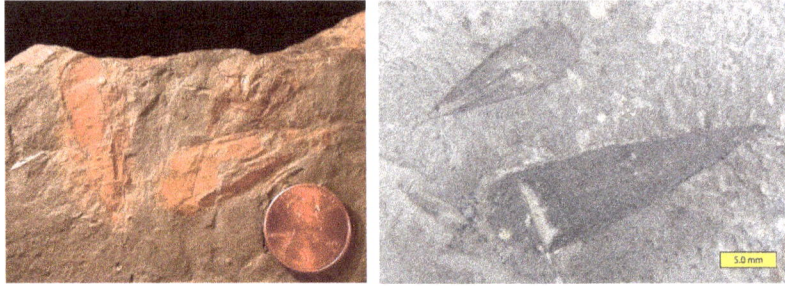

Although apparently extinct, these creatures seem to have risen suddenly and without any ancestors.

9. **Sponges** - are animals of the phylum Porifera, meaning *pore bearer*. They are multicellular organisms that have bodies full of pores and channels allowing water to circulate through them. Sponge bodies consist of jelly-like mesohyl which is the gelatinous matrix within a sponge. Most sponges rely on maintaining a constant water-flow through their bodies to obtain food and oxygen and to remove wastes. These creatures are naturally adapted to the water and they are an example of a working system of parts with various functions. How did evolution accomplish this? How did sponges *evolve* the coordinated ability of various parts to utilize the flow of water in order to *adapt* to the water? All the parts had to have been functioning from the very beginning of their existence, or sponges would not have survived the water environment!

The sponges are amazingly diverse but show no evolution from a common ancestor.

10. **Corals** – are marine **invertebrates** in the class **Anthozoa** of the phylum **Cnidaria** typically living in compact colonies of many identical individual *polyps*. This group includes the important reef builders that inhabit tropical oceans and **secrete** calcium carbonate to form a hard skeleton. Coral fossils are abundant in the Cambrian Period and there is no evidence for their evolution in lower strata.

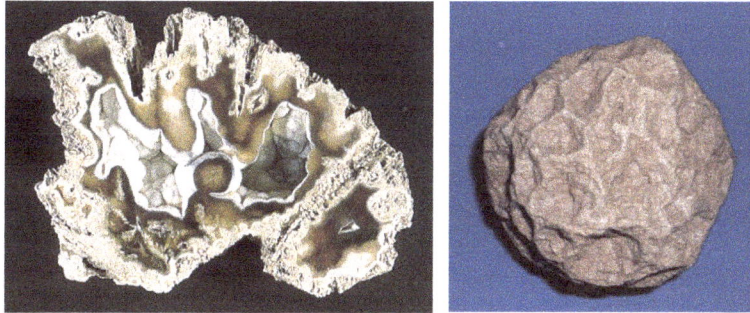

The corals show an amazing amount of variation and are very active today. But they show no evolution from a common ancestor.

11. Radiolaria – include the foraminifera (*protozoa*) and plankton.

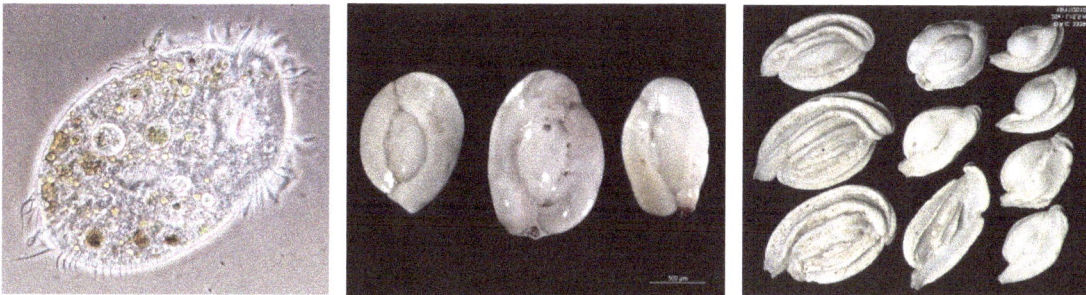

Many varieties of wonderfully preserved protozoa abound in the fossil record. Their evolution certainly would have required hundreds of millions of years to have the kind of diversity present in these fossils. This picture shows the variety of these protozoa that have been found as fossils.

12. *Bryozoa* – are a phylum of aquatic invertebrate animals. They are filter feeders that sieve food particles out of the water using a retractable lophophore, a *crown* of tentacles lined with *cilia*. This kind of complexity certainly must have developed in the layers below the Cambrian. But no ancestors have been found.

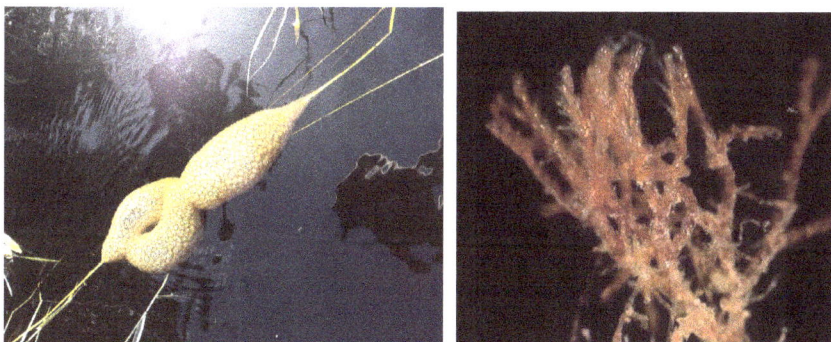

Modern Bryozoa: Incredible variation today among the bryozoa is evident but show no evolution from a common ancestor.

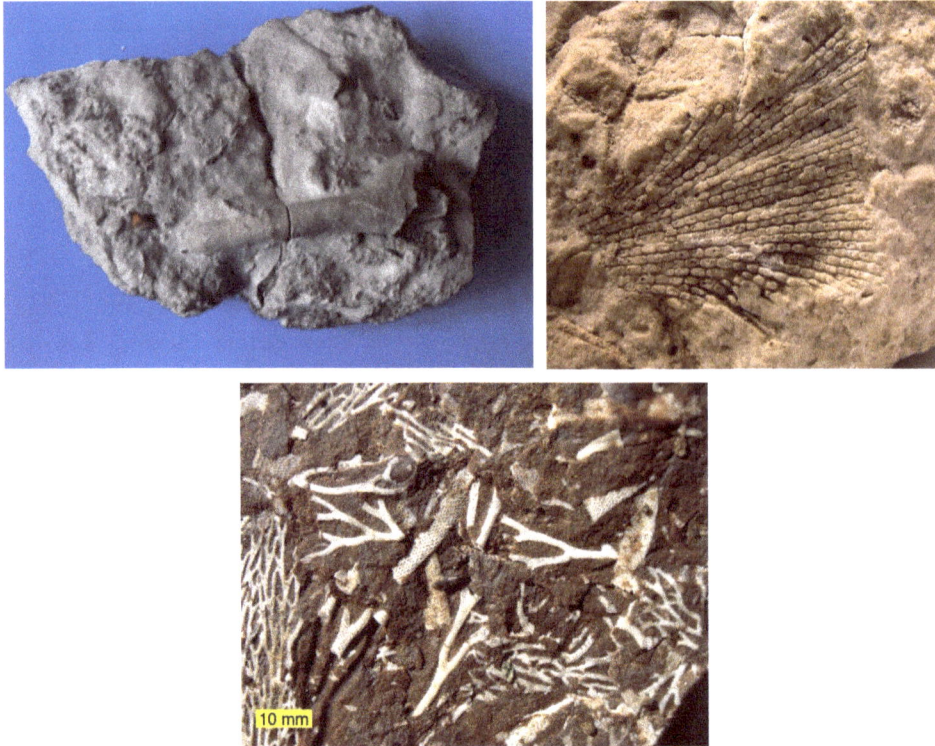

Fossil bryozoan. There is incredible variation among the bryozoa in the fossil record, but there are no documented fossils that have shown the evolution history from a common ancestor.

An artist's rendition of life present during the Cambrian – But where did it come from? How did it evolve? What were the stages of evolution that took place? For the evolutionist, this is all shrouded in mystery. For the creationist, the Book of Genesis explains it quite adequately.

The Cambrian Explosion has generated extensive scientific debate. The seemingly rapid appearance of life in this *primordial strata* was noted as early as the 1840s, and in 1859 Charles Darwin discussed it as one of the main objections that could be made against his theory of evolution by natural selection. The paleontologist Stephen Gould commented, "The fossil record had caused Darwin more grief than joy. Nothing distressed him more than the Cambrian explosion, the coincident appearance of almost all complex organic designs..."[1]

So, is there anything below the rocks labeled as Cambrian Period? In the last 30 years many anomalous fossils have been discovered in what are considered to be older strata below the Cambrian. This stratum of rock has been called the Ediacaran Period. The Ediacaran Period was named after the Ediacara Hills of South Australia where these rocks were first studied. This period was supposed to have lasted around 95 million yeas and represents that period of time immediately preceding the Cambrian Period.

These are some of the Ediacaren fossils that have been discovered. What are they? It's anybody's guess. Some paleontologists have even questioned whether they simply might be some kind of geological impression and not fossils at all. Can these really be the ancestors to the Cambrian animals? There is no unanimous agreement among paleontologists as to what they are. It should be remembered that the Cambrian Period has been drawn at an imaginary line based on the idea that evolution has occurred and that the Cambrian Period represents the lowest forms of life. That raises another question. If we take away this imaginary line, could these Ediacaran fossils simply be part of the Cambrian Period?

Exactly what those fossils are that have been discovered in the Ediacaran has been the subject of very heated debate among paleontologists. These fossils are interesting, but don't really fit into any known **taxa**. Could they be the ancestors of the Cambrian animals?

---

[1]Gould, Stephen Jay. *The Panda's Thumb*, New York: W.W. Norton and Co., 1980. pp. 238-239. Print

Although many paleontologists would like them to be, these fossils hardly come close to being transitional stages of the Cambrian *fauna*. There are no fossils of these strange creatures that show any kind of evolutionary transition to the animals we can clearly identify in the Cambrian. The mystery of the Cambrian Explosion remains a mystery, except for those who trust the Book of Genesis. The evidence is in perfect harmony with what we read in the very first chapter of the Bible.

One example of a fossil that has been identified as Ediacaran by paleontologists is the Sea Pen. Following is a picture of a **Fossil Sea Pen** dated by secular geologists at 560 million years old. This date would put it in the upper Ediacaran, according to uniformitarian dating. Next to the fossil is a living Sea Pen! The living Sea Pen looks exactly like its fossil! There appears to have been absolutely no evolutionary change in the Sea Pen since it first shows up in rocks that are supposedly over 500 million years old.

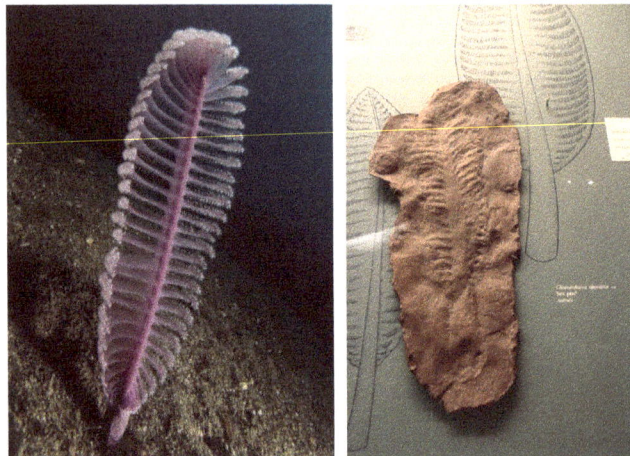

If paleontologists date the fossil at 560 million years old and the Sea Pen is still thriving after all this time, why has evolution not changed this creature? In all that time there must have been environmental challenges that would have caused the Sea Pen to either adapt or go extinct. And where are the transitional fossils that would show how the Sea Pen developed? Perhaps, in reality, the Sea Pen is really not that old!

According to Genesis chapter one, plants were the first living things to be created, then sea creatures and then birds. But the order in the fossil record from an evolutionist perspective *starts* with sea creatures, as they appear to be located at or near the bottom of the fossil record. Early evolutionists interpreted this to mean that the first life that developed on Earth must have been sea life, because their fossils were found in bottom rock layers. Because the Genesis Flood had been rejected as a legitimate explanation for this order, 19th century geologists made a serious error. They reasoned that the rock layers represented living environments that had recorded perhaps millions of years of living things. Those creatures that lived a long time ago were buried on the bottom. But is that what the rock layers represent?

If we look at things from a Scriptural viewpoint, the rock layers could represent a general order of the Genesis Flood burial. We find sea creatures mostly in the bottom rock layers, not because they were the first to come into existence, but as the first casualties of the Flood. As the Flood began and the fountains of the great deep burst open, the ocean bottoms would have been torn up and buried first. Along with these deposits would have been buried massive amounts of sea life. This would have been followed by land deposits with the remains of land and plant creatures. It is a whole different way of looking at the rock layers, isn't it?

The significance of the Cambrian Explosion from a Genesis perspective is that it represents well-designed life that seems to have sprung up suddenly and exhibited rapid diversification from the beginning. Genesis 1 teaches that life was created instantaneously from nothing followed by rapid diversification into many varieties soon following. What we have discovered in the rocks is that they explain what God did in Genesis chapter 1 – amazing life with an ability to reproduce after its kinds.

That brings us to the next lesson in our study of fossils, the Genesis Flood. In the next two lessons we will examine the cause and mechanisms of the Genesis Flood and what it has to do with the billions of fossils we have been discovering since the 1800s. What do the fossils really tell us?

Such exquisitely preserved delicate parts of living things such as intricate wing detail of insects, eggs, delicate bones and eye sockets, delicate legs, fine bones and the markings on insects really require a different explanation than the uniformitarian one. Such fossil details are not the exception, but the rule in the fossil record. They could not have been preserved in huge numbers if nature had taken its usual course of decay and disarticulation. The animals represented in the fossil record must have been buried rapidly in some kind of catastrophic watery burial.

Activity 2

Take out the invertebrate fossils from your kit. Using a hand magnifier, observe each fossil and write out what you see. Keep your notes in your workbook or lab book.

Please take Quiz #2, Appendix B

Lesson 3 – The Genesis Flood, Part I

Part I, The Cause of the Flood

Ivan Ayvazovsky, Noah's Flood – this painting captures the purpose of the Genesis Flood. It was a divine judgment on sinful mankind that had corrupted the Earth. This judgment is depicted by the Book of Genesis as violent, overwhelming, complete, and global.

Words to know: *sin (sinfulness) wickedness evil disobedience canon revelation Deist*

Introduction

Modern science has rejected miracles and God's involvement in His creation as being unscientific. In modern science naturalistic causes alone must be able to explain the physical data. The thought of a God initiating a geological event such as Noah's Flood is simply myth and certainly unscientific. Science rejects the Genesis Flood! It is nothing more than a myth similar to other myths like those of other civilizations. This has produced blindness to the most horrific geological event in the history of man – and because of this, modern geologists have misinterpreted the geology of the Earth!

The Book of Genesis, chapters 6-9, teaches in no uncertain terms that around 4,500 years ago there was a global flood. It is clear from these passages of Scripture that this event was the greatest geological event in Earth's history, except, perhaps, for the six-day creation. Why did it come on the Earth? What was the cause? Was God actually directly involved in bringing it about?

The cause of the global flood of Genesis
In **Genesis chapter 6:5-7,** we are told of the cause of the Flood:
a) The *wickedness* of man was great
b) Every intent of the thoughts of the heart of man was only *evil* continually
c) It grieved God to such an extent that He decided to wipe man out and start fresh with Noah and his family

The Genesis Flood was no natural event. It was brought on by God Himself because of the wickedness of man. The lesson of the historical Flood is clear. God holds man accountable for the choices he makes. The Flood was a judgment by God for the wickedness of man and the intent of his heart – only evil continually.

One of Charles Lyell's main goals in the 1800s was to eliminate the idea of a global flood and the Book of Genesis from modern geology. Why the Flood? The underlying reason is clear. If the Flood was a judgment from God on the **sinfulness** of man, then it is one of the clearest historical events that teaches man's accountability to the God who created him. It follows that it would also be one of the most attacked teachings of Scripture. History teaches us that man spends much energy trying to escape the thought of a judgment and the consequences. The historical account of the Flood in Genesis tells us that every intent of man's heart was on evil continually. Perhaps he will do everything he can to escape the idea of his accountability to his Maker and His judgment.

This idea is repulsive to The Enlightenment concept of man. The enlightened man, as the highest creature in the various divisions of life, was capable of great thinking and unlimited goodness. The idea that man is basically good and that he only needs the right education, came out of this period of time.

At this point many people wonder if it was God's fault that He made man that way. This is not a correct understanding. Genesis 1:31 says, "And God saw all that He had made, and behold, it was very good." There was no sin, no evil, and no blemish. Everything was in perfect harmony and unspoiled. It is in the 3rd chapter of Genesis that we see what actually happened. Adam and Eve disobeyed God. They chose to listen to another voice, a lie, and consequently brought sin and death on the Earth and to their descendants. When geologists of the 19th century chose to cut out Genesis from their understanding of Earth history, they cut out the very reason for a global destructive flood. The cutting out of the morality of the Bible soon followed in the late 19th Century. The 21st century is now reaping the consequences of this abandonment of Scripture during The Enlightenment. Not only are we left with a naturalistic explanation of the heavens and the Earth, but we are left without an explanation for evil and sin.

Everyone regardless of whether he or she is an evolutionist or theist is aware of the concept of evil. And all acknowledge to some degree that people who practice evil should be punished in some way. But what is the universal standard that we are going to use to define what is evil? Without the Scriptures, the definition of evil can vary from culture to culture. Whose standard should we use? There is no adequate or universal definition of evil. The Scriptures make it clear; evil does exist, and it is defined as **disobedience** to God and His laws.

Since the **Deists** of the 1800's rejected the Scriptures as historical they were left with an arbitrary standard. To them, the idea of a global flood that was brought on by God for the wickedness of man seemed barbaric. This is one of the reasons the Deists tried to relegate the Bible to myth. Even today many people cannot conceive of a God who would do such a thing. The Genesis Flood is a clear historical account and demonstration of what evil is and what its consequences should be. No wonder geology is such a battle ground for the church today. If man could eliminate the Flood from any scientific consideration, then the message of the Genesis Flood would be lost.   This is why it is critical that we understand the importance of the Scriptures and accept its historical account of the Flood.

The importance of the Hebrew Scriptures
The word *scripture* is from the Greek language and means, *the writings*. In Luke 24:25-27 Jesus rebukes his disciples as foolish for not believing in all that is written in the Scriptures.

Luke 24:25, 27 - "O foolish men and slow of heart to believe in all that the prophets have spoken!...Then beginning with Moses and with all the prophets, He (Jesus) explained to them the things concerning Himself in all the Scriptures." When He did this, He was referencing the Hebrew Scriptures. The Apostle Peter includes the writings of Paul (2 Peter 3:15-16) as Scripture. The church has universally accepted certain writings over others by standards that mark them as authoritative and doctrinal. This collection of writings by those who were intimately associated with the Lord and His mission has been called the *Canon*. The word is from the Latin for *the rule*. The Canon of Scripture is the rule which Christians follow. The Canon is a closed issue and has been demonstrated to be authentic and God-ordained. To this day not one doctrinal error, contradiction or historical error has ever been found in the Scriptures, even though they have been copied thousands of times. They are the most documented and scrutinized pieces of historical literature in existence today!

The Scriptures or Bible as most people call them today, are also called **revelation**. Revelation is a unique characteristic of the Scriptures. Many have tried to duplicate this characteristic through visions, crystal balls, palm reading, astrology, predictions of the future, and palm reading, but they have failed miserably. Man wants to be in control of his own destiny and will look to all kinds of things to help him to do that; everything except the Bible. We are explicitly warned against attempting to discover the future or of hidden things through anything other than the Scriptures or God's prophets. This is recorded in Deuteronomy 18:9-22. God's people were not to have anything to do with:

a) The detestable things of the nations
b) Idolatrous customs of the nations
c) Divination; the use of physical objects to aid in discerning the future or the past
d) Witchcraft; the use of spells and incantations to aid in discerning the future or the past
e) Omens or sorcery
f) Those who cast spells or those who call themselves mediums or spiritists or necromancers (those who contact the dead to gain insight into life or the unknown). The one to whom Israel was to listen to was the prophet that God would raise up to speak what God had commanded. In fact, if a person claimed to be a prophet and was not, he was to be stoned! The Scriptures are absolutely crucial to understanding the past, the present and the future. It is no wonder that the serpent from the

beginning has chosen to undermine what God has spoken. The Scriptures are the keys to the whole creation-evolution conflict.

Biblical revelation

Revelation is a central doctrine to Christianity. It is the unveiling of God's will and purpose throughout history. History began in Genesis 1, "In the beginning…." There was no prehistory or prehistoric past beyond this first verse of Genesis 1. God revealed enough of history to let us know about our origin, our purpose and our problem. He also revealed enough about Himself in order for us to find Him and turn to Him. The creation or nature on the other hand is a witness to God's Being and His works. Romans chapter 1:18-20 tells us that we can understand certain things about God through what has been made:

a) His invisible attributes

b) His eternal power

c) His divine nature

Some have called nature the 67th Book of the Bible, but this is not an accurate statement. The Scriptures are *writings* that reveal history; the creation is a *work* of God and a *witness* to that work. Nature only testifies to the fact of God and His work in creation, it does not reveal the written history of God's plan and work.

What effect would a global flood have had on the Earth and what did it have to do with fossils? We will explore the answer to this in Lesson 4.

Activity 3

Using a concordance of the Bible, find 25 passages that describe what evil, sin and wickedness are. Write out these passages and the specific things that make them evil, wicked and sinful.

Please take Quiz #3, Appendix B

Lesson 4 – The Genesis Flood, Part II

The Mechanisms for a Global Flood and the Fossil Evidence

The tremendous tectonic activity of the Genesis Flood would have torn up huge amounts of earth and soil, transported them great distances complete with billions of dead things in mud strata all over the Earth.

Words to know: conglomerate paleontology tectonic mechanism velocity phenomena hydrological

At least from the point of view of modern geologists – *Scientists have the proof for evolution in the fossils. Creationists only have words. What do Christians have as evidence for creation and a flood?* They then mockingly say, "They have their Bibles!" In other words, the implication is that creationists do not have any scientific evidence for their beliefs other than fairy tales.

We have the fossils.
We win. 🐟

The evidence for the Genesis Flood is abundantly found in the fossils. Believers in the Bible must be willing to spend the time to understand this physical evidence and its significance in explaining the Flood. The Book of Genesis adequately accounts for the fossil evidence. We simply have to invest the energy and time to learn and understand how the fossils fit into the Flood account.

The Book of Genesis teaches that the space, Earth, the sun, moon, stars, and all living things were created in six literal days. Then shortly after this, Adam and Eve disobeyed God and brought sin into the world. Evil and wickedness began to grow to such an extent that God stepped in and judged it with a global flood. In Genesis chapter 7 we are told in no uncertain terms that a global flood took place during the days of Noah, roughly 1,600 years after the creation of Adam. Is there any physical evidence to support this Biblical view of Earth history? Actually, there is! Evidence for the Genesis Flood is provided from both the field of geology and from the field of **paleontology**. Since our study is mainly focused on fossils, we will examine evidence from paleontology. By paleontology, I mean the study of fossils. The modern discipline of paleontology portrays a naturalistic explanation for the origin and nature of the fossil record in the rock layers. The Genesis Flood gives a completely different explanation. Let's look at some of the physical, fossil evidence that has been uncovered.

Consider this excerpt from the article, "World's Largest Dinosaur Graveyard Linked to Mass Death."

> The dinosaurs may have been part of a mass die-off resulting from a monster storm, comparable to today's hurricanes, which struck what was then a coastal area...thousands died in the flood... The likely culprit in this scenario was a catastrophic storm, which could quickly have made the waters rise up as high as 12 to 15 feet, if experiences with modern floodplains are any guide. The flooding could have reached more than 60 miles from the shoreline, Eberth told LiveScience. "The landscape basically just drowns... It's unlikely that these

animals could tread water for very long, so the scale of the carnage must have been breathtaking," Eberth said.[3]

A "monster storm," "thousands died," "a catastrophic storm," "reached more than 60 miles," and "the scale of carnage" are statements that don't describe an ordinary "storm." Why not just call it, the Genesis Flood?

Bone Beds or Graveyard Fossils

These landforms, called bone beds or fossil graveyards, are treasure troves of fossil remains. They are called graveyard fossils because they appear to be a mass burial of hundreds of thousands of plants and animals - many different animals from totally different environments in some cases. That is not what we observe in the natural course of life and death today. Mass burial may occasionally happen on a local level, but it is not the normal flow of life and death. Usually individual plants and animals die and decay. Their remains are rarely, if ever, preserved. In fact, it is so rare that scientists have trouble explaining just how something becomes fossilized. Let's examine some of these graveyard fossils. Read the following excerpts from articles published at Creation Ministries International (Creation.com).

From Tas Walker:

> Researchers from the USA and Chile reported, in November 2011, a remarkable bone bed on the west coast of northern Chile near the port city of Caldera, about 440 miles north of the capital, Santiago. Excavations uncovered the remains of some 80 baleen whales of which more than 20 specimens were complete. They also found other kinds of marine mammals including an extinct dolphin with tusks and a sperm whale... The puzzle of how these marine creatures died has caught news headlines with one reporting "Fossil Bonanza Poses Mystery". Another asked, 'How did 75 whales end up in the desert?"... The field evidence for large-scale catastrophe is overwhelming as these research scientists have reported... The sandstone strata containing the whale fossils are contained within a local area called the Caldera basin... the characteristics of the sediments in these basins and the abundant fossils contained in them indicate that deposition took place during a period of rapid and major coastal subsidence

---

[3] Charles Q. Choi, "World's Largest Dinosaur Graveyard Linked to Mass Death." Livescience.com. 18/2/2014.

(subsidence is the motion of a surface usually, the Earth's surface, as it shifts downward relative to a datum such as sea-level. The opposite of subsidence is uplift, which results in an increase in elevation.) ... And major coastal subsidence explains the rapid burial of the whales and other creatures because rapid burial was needed soon after death to preserve the fossils.[4]

Fossil whale bones: Normally, fossil bones are found in a disarticulated (disjointed) state, like the pictures above. Very rarely are complete, intact skeletons found. That is why the find mentioned in the previous article is so unusual.

From Andrew Snelling:

(A)t Partridge Point are outcrops of a fossil graveyard, one of many fossil deposits found around the world that are significant as examples of the catastrophism during the Flood. ... here we also see, thrown together with these crinoid columnals, pieces of 'lace coral', brachiopod shells, and solitary corals. The limestone that now entombs these remains cannot be where these creatures once lived, because they are not found here in their living positions. The solitary coral, for example, is not seen here attached to the sea bottom, with a distinct hard surface visible in the rock mass, but was completely enclosed in what originally was a soft lime mud which only became rock hard after burying the coral. There is, therefore, only one conclusion which makes sense of the evidence—these sea creatures were buried together suddenly when overwhelmed, carried, and then dumped by moving water filled with lime muds. That's why geologists call this a fossil graveyard. However, unlike a human graveyard today, where the individual graves are neatly arranged, these fossils in these graveyards are all jumbled—tossed together and buried haphazardly in sediments laid down by moving water. But the scale is also impressive. Only a tiny portion of the Thunder Bay Limestone is exposed at Partridge Point, whereas these rock layers extend sideways for several hundred miles in each of two directions across what is known as the Michigan Basin. At Partridge Point one can see countless thousands of fossilized sea creatures' remains. But one's mind is quickly overwhelmed trying to comprehend the countless billions of fossils that must therefore have been buried in these rock layers underneath many hundreds of square miles of Michigan! And this is only one of the fossil graveyards found in Michigan. In fact, similar fossil graveyards

---

[4] Walker, Tas. "80 Whales buried mysteriously in Chilean Desert: Marine graveyard is evidence for Noah's Flood." Creation.com: 01 December 2011. Digital.

are found in many places on every continent all around the globe— "billions of dead things (fossils) buried in rock layers laid down by water all over the Earth." [5]

This is exactly the evidence we would expect to find based on what the Bible says about the Genesis Flood.

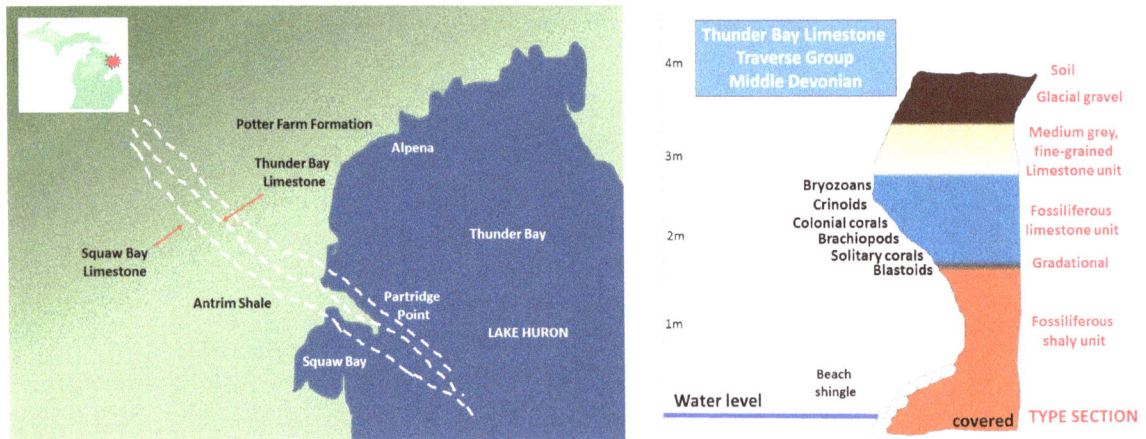

Location map for the Alpena area on the Lake Huron coastline of eastern Lower Michigan (USA), showing Partridge Point and the local extent of the Thunder Bay Limestone. Idealized composite diagram depicting the type or reference section of the rock types and layers making up the Thunder Bay Limestone along the Lake Huron shoreline at Partridge Point. Most fossils found in the shingles of limestone along the beaches have eroded from the fossiliferous limestone unit, and some of these fossils are listed there.

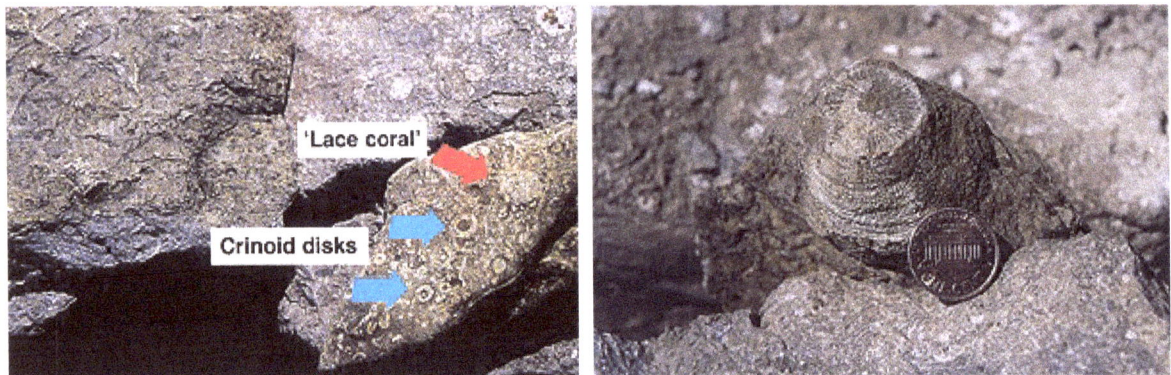

Crinoid columnals or disks from the stalks (end-on and side-on views) scattered haphazardly through the Thunder Bay Limestone. Also shown is some 'lace coral' (a bryozoan). A solitary coral (U.S. penny for scale) surrounded uniformly by what was originally lime mud but now limestone.

[5] Snelling, Andrew. "Thundering Burial." Creation 20(3):38–41 June 1998. Accessed 17/2/2014. Digital.

Again, from Tas Walker:

An international team of scientists have uncovered graphic evidence of the deadly terror unleashed on a herd of dinosaurs as they were buried under sediment by the rising waters of Noah's Flood in western Inner Mongolia. Dinosaur bones were first discovered at the site, located at the base of a small hill in the Gobi Desert, in 1978 by a Chinese geologist. After about 20 years, a team of Chinese and Japanese scientists recovered the first skeletons, which they named Sinornithomimus, meaning "Chinese bird mimic". A few years later in 2001, the international team excavated the remains of more than 25 dinosaurs, creating a large quarry in the process as they as they followed the skeletons into the base of the hill... Most of the dinosaurs were buried in a life-like crouching posture and, even more surprisingly, the limbs of the dinosaurs were plunging down into the underlying mud as deep as 40 cm. Their hind legs were often still bent indicating that they were struggling to escape. Two of the skeletons were found one right over the other where they apparently fell. This fossil find captures in stone how the dinosaurs perished when they became mired in the mud. Not only was the thick under layer of sediment recently deposited, but the overlying sediments were deposited soon after the animals were trapped, burying the animals before their soft parts had a chance to rot away. Nearly all the fossil bones were surrounded by a drab, blue-gray halo indicating how far the soft tissue extended, and that the carcasses had decomposed after being buried, not before. In addition, gastroliths (stomach stones) were found in the fossilized ribcages of some animals, as well as carbonized stomach contents. So promptly were the animals buried that the delicate bones in the eye (sclerotic rings) of some animals were preserved. The team interpreted the site as a "catastrophic miring of an immature herd... One problem that the paleontologists encountered is that according to uniformitarianism the fossils layers preserve a living environment that existed at that time. Therefore, the team was surprised that the dinosaurs consisted only of juveniles without any adults or hatchlings present. However, that is perfectly understandable in the Flood catastrophe when animals were fleeing. You would expect the hatchlings to have already perished and the adults to have fled and abandoned the youngsters. In scientific circles these sorts of anomalies are never reported as a problem. Rather, the paleontologists reported this unexpected result as simply that dinosaurs were left to fend for themselves in juvenile herds while the mature adults were occupied elsewhere with parental care of eggs and hatchlings. What an amazing story. Another problem for the team was the thickness of the mud in which the dinosaurs were trapped. They suggested the area was a low energy lake environment, which is the standard interpretation that uniformitarians invoke to explain muddy sediments... Another problem is that the team found mudcracks on the mud, which they also interpreted as indicators of drought. Mudcracks form when mud emerges from the water and has dried for a day or so. How could the mudcracks form on the mud surface if it was in deeper water? This array of evidence that conflicted with their expectations puzzled the team and they once again presented the results as an "exceptional" discovery. However, the thick mud deposit, rapid sedimentation and catastrophic entrapment of the animals are easily explained by the Flood catastrophe. And mud does not need to be exposed above water for mud cracks to form. Shrinkage cracks will form in situ once the overlying sediments have been deposited and the water within the mud is expelled and the mud contracts... Here in Inner Mongolia in the middle of Asia the historical reality of Noah's Flood explains the new dinosaur finds

47

elegantly. The herd of dinosaurs was a casualty of the enormous watery catastrophe that engulfed the region during the Flood. They were overwhelmed during the first half of the catastrophe as the waters were rising on the earth, while air-breathing, land-dwelling animals were still alive. Sediment continued to accumulate on the continent during this Inundatory stage as the waters continued to rise. Then, when the waters receded from the continents they eroded some of the overlying material, shaping the landscape, and leaving occasional erosional remnants, such as the small hill where the geologists were able to excavate this dinosaur graveyard.[6]

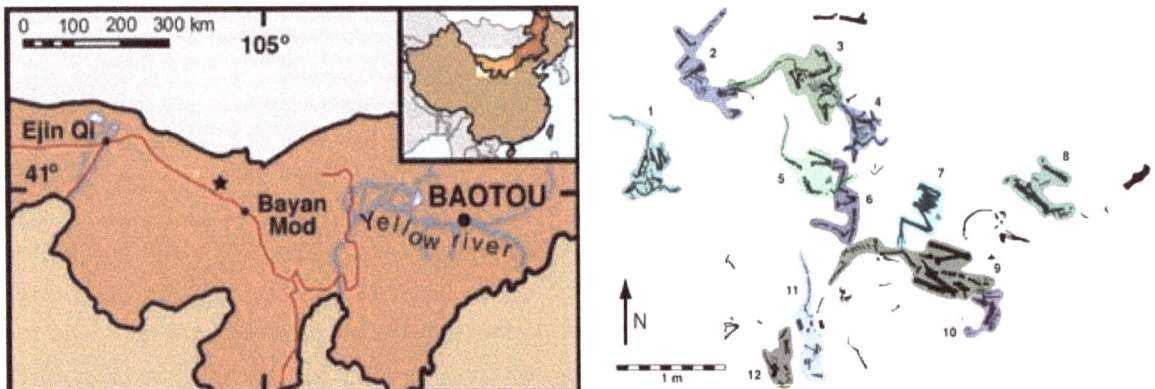

Location of the fossil site in Inner Mongolia, China - Map of some dinosaur remains at the site in Inner Mongolia. Note that the skeletal parts have generally remained together indicating that the animals were buried before their remains disintegrated – obviously indicating some kind of catastrophic burial.

Evidence for a global catastrophe abounds from all over the world. The previous articles just begin to scratch the surface of the geological and fossil evidence for the global flood of Genesis. In order for the billions of fossils to have been buried such that they would have escaped decay, nothing short of a global flood could explain it. The fossils in the many layers of rock all around the world tell us of just such an event – the global, catastrophic flood of Genesis.

Geologists tell us today that fossil preservation is an extremely rare event. The Genesis Flood would have indeed been that rare event! Nothing like it has occurred since. There must be a record of the geological upheaval that most certainly would have taken place with an event of this magnitude. So let's examine what might have taken place geologically during the Genesis Flood. Does the Genesis account give us any hint of the geology of the Flood?

---

6 Walker, Tas. "Dinosaur herd buried in Noah's Flood in Inner Mongolia, China," Creation.com: 14 April 2009. Digital.

The geological mechanisms of the Genesis Flood

What is a *mechanism*? It is the means by which an effect is produced, or a purpose is accomplished. Does the account of the Flood given in Genesis describe any mechanisms involved in this global event?

One of the great claims of the Deists of the 1800s was that in order for things to be scientific, there could be no intervention by a God into the geological history of the Earth. But of course, that proposal is in itself a belief. Science cannot prove one way or another whether God was involved or not. But the Bible is an historical account and as such we must consider that account, if we are going to have an adequate view of the history of the Earth. If the global flood of Genesis was an actual historical event, then so is its record of the involvement of God. The Bible definitely records that God was involved in the Flood – from first to last. So, we have an historical account that has never been disproved, only rejected based on a competing worldview, not scientific evidence.

The mechanisms of the Genesis Flood

From the Genesis account we can see four mechanisms which would have had significant geological effects:

1. God initiated the Flood. An initial period of 40 days and nights of rain which must have been furious.

2. The fountains of the great deep burst open; the beginning of the Flood that produced *tectonic* upheavals and a source of water along with the opening of the floodgates of the skies.

3. Rising water until the mountains that existed then were covered by $22\frac{1}{2}$ feet.

4. Receding water until land was visible; this most likely occurred in a combination of sheets and channels.

When most people think of Noah's flood, they think of a rainstorm that lasted forty days and forty nights. In their minds, that is usually the extent of the Genesis Flood. I know that this is exactly what I thought as a kid. Noah's flood was a story, nothing more. However, if the whole story (Genesis 6-9) is read in its entirety, it will become apparent that there was so much more that went on geologically. God Himself was involved in this geology from the very beginning of the Flood.

I have taken these mechanisms and put them into a visual diagram of events. What this diagram shows are the major geological events of the Genesis Flood as outlined in Genesis 7-9.

Geological Events of the Genesis Flood
Genesis 7-9

Time Duration	Stage of the Flood	Phase of the Flood
371 days		Water in channels
	Receding Stage	Water in Sheets
150 days		The Peak
	Flooding Stage	Water Rising
40 days		Flood Beginning

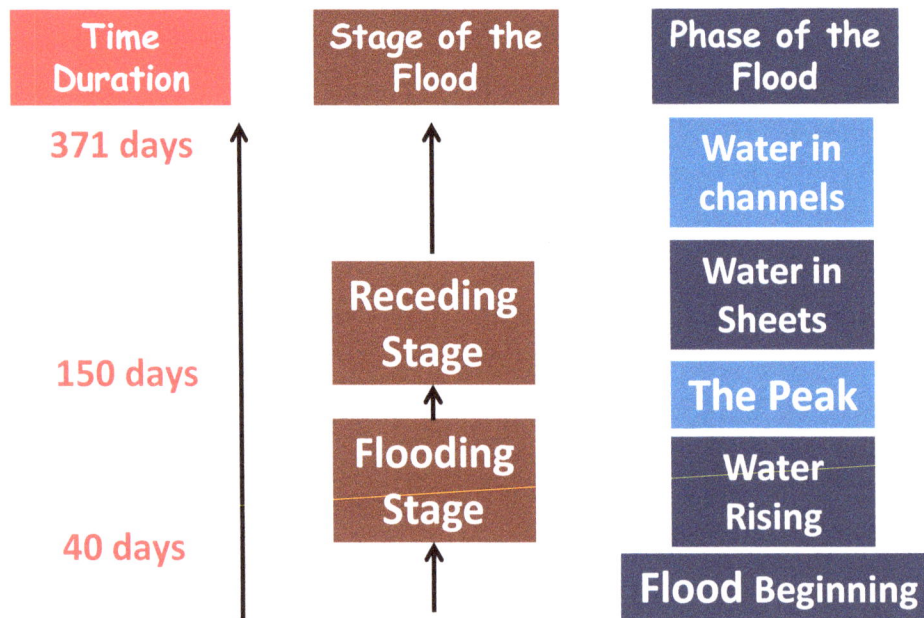

First, notice the length of the Genesis Flood – 371 days! The days of the Flood, according to Genesis 7-9 are further divided into two distinct Flood Stages with the 150th day being the apex of the Flooding Stage.

Second, notice the Receding Stage (also called the Retreating Stage) of the Flood. According to Genesis 7-9, the Receding Stage of the Flood lasted for an additional 221 days! That makes the Flood event significantly longer than the typical 40 days and 40 nights of rain that most people think of.

Now, although the Bible is not an academic textbook on geology, it is an historical account that provides us with a geological framework. And this framework can provide us with a mental picture of what might have taken place. It provides us with a framework to talk about what might have happened geologically during this global catastrophic event. Furthermore, using this framework as our guide, it is amazing how the rock record does match up with this in many places around the world.

1st stage of the Flood – the Flooding Stage

Where did all the water come from for the flood? Geologists tell us that if all the mountains were leveled off, the water in the ocean would cover the Earth to a depth of five miles. If the present ocean existed during the Flood, we have to explain not only where the "other" water came from, but where the then existing Flood waters went to.

In Genesis 1 the account tells us that the originally created waters were gathered into one place and that dry land then appeared.   So, at the very least, the pre-flood Earth looked a whole lot different than it does today. This would have been true for the amount of visible water as well. The Genesis account also tells us in chapter two that the way the Garden was watered was from a mist that used to rise and water the whole garden. Without discussing the geographical boundaries of the Garden of Eden at this point, we can see that there must have been some kind of underground reservoirs or aquifers that may have stored a lot of water. We do know there are presently many aquifers under much of the Earth. Could these be remnants of the water that was once stored under the Earth? Below I have given two pictures of what is considered to be the largest aquifer in the world. Could this be one of the remnants of the water supply for the Flood?

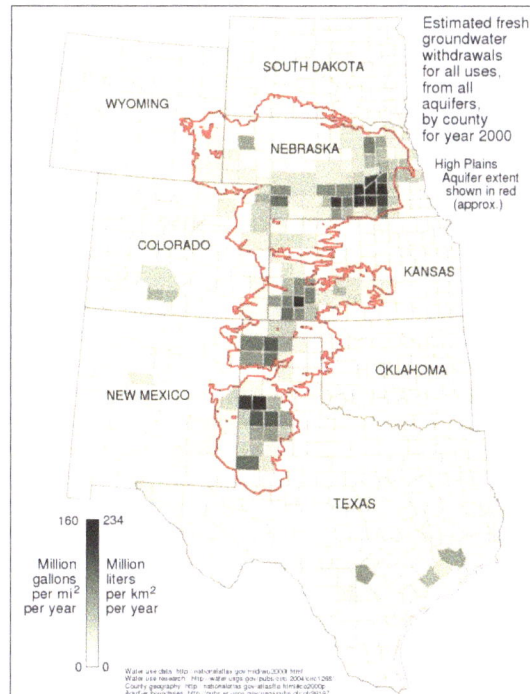

The red outline on the map is the approximate range of the largest fresh water aquifer in the world, the Ogallala Aquifer located in the Great Plains of the United States.

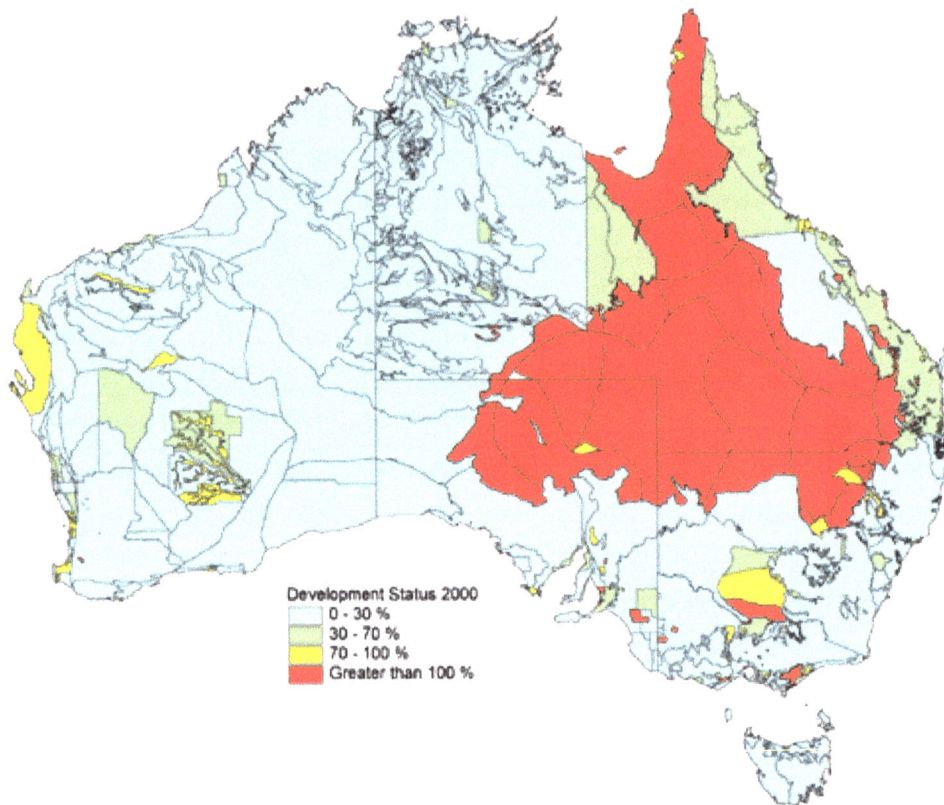

The area in red is another immense aquifer located under eastern Australia, called the Great Artesian Basin.

These present-day aquifers may in fact be the leftovers of what were once huge amounts of water stored beneath the Earth. The oceans of today may be the water that came from these aquifers. The event that describes the release of this water is found in Genesis 7:11 at the outset of the Flood. "...(T)he fountains of the great deep burst open". Walt Brown (Ph.D. from MIT, director of the Center for Scientific Creation) has used a term he calls *the hydroplate theory to* describe what might have happened in the initial phase of the Flood. This idea would incorporate huge amounts of water that were stored beneath the Earth shooting up through the crust and into the atmosphere to produce rain that had never been experienced before, nor ever will be.

Once the Earth was covered during the first 150 days of the Flood, the big question is, "Where did all the water go?" That brings us to the next stage of the Flood.

2nd stage of the Flood – the Receding Stage

So far, the picture is one of a flooded Earth; no land visible anywhere, not even mountains. Everything was covered all over the globe. How did the water recede and where did it go?

In Psalm 104:7-9, according to the Biblical account of the Flood, God rebuked the waters so that they fled. He caused the mountains to rise and the valleys to sink. Rapidly rising mountains such as the Rocky Mountain chain would have helped to create rapidly receding water. Fast-moving water can produce a tremendous amount of geological change in a short period of time as is witnessed by tsunamis today. Our diagram above of The Geological Events of the Genesis Flood shows that there might have been two separate *hydrological* events involved in this process – a sheet-flow stage (water in sheets) and a channelized stage (water in channels). Both of these geological events would have produced planed surfaces and great canyons. The rapidly receding water flowing off of rapidly rising mountains would have moved huge amounts of freshly-laid sediments with tremendous *velocity*. More resistant sediments would have remained to leave structures like those found in Monument Valley in Arizona.

Monument Valley in Arizona and Utah that is essentially a planed land surface with remnants of the sediments that used to cover this area.

The Grand Canyon is essentially a planed surface with a huge gash in it. The layers of sediment were laid down in the Flooding Stage of the Flood. In the Receding Stage of the Flood the sheet-flow stage of the Flood would have planed the freshly-laid sediments followed by the channelized stage which would have cut huge gashes in the remaining sediments.

Just how big are the oceans of today? Geologists tell us that water covers about 72% of the Earth's surface. Only 5% of the world's oceans have been explored! Most of this water is what remains of the global flood of Genesis 6-9.

I believe these short Scriptures in Genesis 7-9 and in Psalm 104:5-9 describe a historical and geological framework with enough information to interpret the geological **phenomena** we observe today.

If the layers of rock which contain billions of fossils are actually evidence for a global catastrophe, then why do modern geologists arrange the same fossils we see into a "time column"? The main reason lies in the late 18th century when the Genesis Flood was rejected as a legitimate explanation of the rock layers. As such, an alternate explanation had to be developed: a naturalistic one, one that would reflect this new geology. And thus, the journey of reinventing the meaning of the rock layers began. In our next lesson we will examine the structure and meaning of the modern Geologic Time Table and how it influences the interpretation of the observable rock layers.

Activity 4a

In preparation for our next lesson on the Geologic Time Table, I would like you to use an internet resource. Go to Creation.com. In the search box type, *geologic time table*. Choose five articles and read them. Take notes on your reading, noting the things that particularly strike you about the Geologic Time Table.

Activity 4b

Draw a picture of what you think it might have looked like, when *the fountains of the great deep burst open*.

Activity 4c

In your kit, you will find a collection of vertebrate, invertebrate and plant fossils. Look at each one closely and describe a set of circumstances under which these plants and animals could have been preserved.

Please take Quiz #4, Appendix B

Lesson 5 – Understanding the Geologic Time Table

Words to know: eon era epoch period interpretation masquerading ecological zone idealized atheism correlation fossil succession

The Geologic Column (also called *The Geologic Time Table* or *The Column*)

Eon	Era	Period		Epoch		Development of Plants and Animals
						Time Units of the Geologic Scale
Phanerozoic	Cenozoic	Quaternary		Holocene	0.01	Humans develop.
				Pleistocene	1.6	
		Tertiary		Pliocene	5.3	"Age of Mammals."
				Miocene	23.7	
				Oligocene	36.6	
				Eocene	57.8	
				Paleocene	96.4	Extinction of dinosaurs and many other species
	Mesozoic	Cretaceous 144		Age of Reptiles		
		Jurassic 206				First birds. First flowering plants. Dinosaurs dominant.
		Triassic 245				
	Paleozoic	Permian 286		Age of Amphibians		Extinction of trilobites and many other marine animals. First reptiles. Large coal swamps. Amphibians abundant.
		Carbonifer.	Pennsylvanian 320			
			Mississippian 360			
		Devonian 408		Age of Fishes		First fishes. Trilobites prominent. First organisms with shells.
		Silurian 438				
		Ordovician 505		Age of Invertebrates		
		Cambrian 570				First multi-celled organisms.
	Proterozoic	Collectively called Precambrian, comprises about 87% of the geologic time scale.				First one-celled organisms.
		2500				
	Archean					Age of oldest rocks.
		3800				
	Hadean					Origin of the earth.
		4600				

Above is pictured a typical Geologic Column or Geologic Timetable of today. The picture conveys a nice, well-ordered arrangement of the history of the Earth and its evolution of living things. But how much of it is actually a true, scientific, and historical representation of Earth's actual history?

Most people in our Western culture now believe that the history of the Earth as told in the Book of Genesis is just a story with no scientific foundation whatsoever. This is a shame. There is plenty of evidence. What people have failed to grasp is that what we have been taught is an *interpretation* of the evidence *masquerading* as science. And this interpretation is now the definitive story of Earth history. One of the teaching tools used to tell this story is the Geologic Column or Geologic Timetable (*The Column*). It is the *Bible* of modern geology. Everything must line up with this standard to be considered scientific geology.

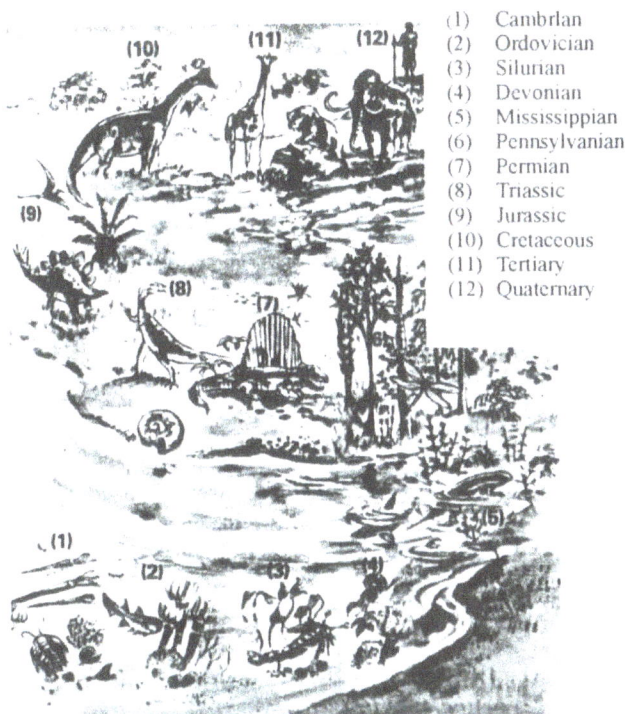

(1) Cambrian
(2) Ordovician
(3) Silurian
(4) Devonian
(5) Mississippian
(6) Pennsylvanian
(7) Permian
(8) Triassic
(9) Jurassic
(10) Cretaceous
(11) Tertiary
(12) Quaternary

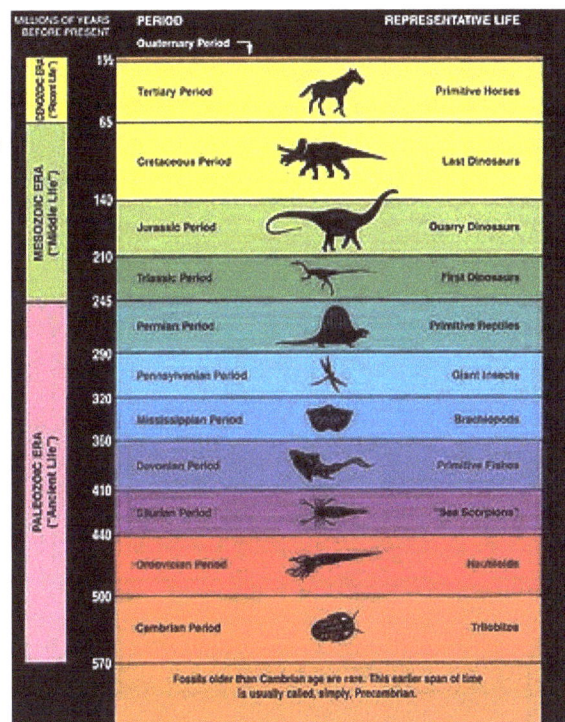

Above are two illustrations that represent two entirely different interpretations of the rock layers. The picture on the left is based on Genesis 7-9. The Flood was a year-long event with devastating consequences. The picture on the right is the evolutionary uniformitarian view of the rock layers.

The pictures above represent two different ways of looking at the rock layers. The one on the left represents the Genesis Flood view in which living creatures at the time of the Genesis Flood were buried by the rapidly advancing flood waters according to *ecological zones.* In other words, sea creatures would have been buried first as they were the first things to be torn up by the Flood, followed by land creatures and finally man. It is a general order produced by the global Flood.

The picture on the right represents rock layers laid down according to a long evolutionary history of life - the layers representing different environments that happened to be preserved throughout hundreds of millions of years of evolutionary history. What is surprising to most people is that there is no place on Earth where the *idealized* conception of the rock layers depicted above can be found. The fossil record is rather a jumbled mess of fossils and sedimentary rocks preserving hundreds of anomalies or exceptions to the concept.

The Column is arranged into time periods called *Eons*, *Eras*, *Periods* and *Epochs* with assigned durations and major geologic events in Earth history. But is it a discovery of scientific fact and the result of scientific investigation of actual history or is it a philosophical attempt to frame a totally different view of Earth history as opposed to that of the Scriptures? The history of The Enlightenment, which gave us modern geology, would show us that modern geology is mostly a philosophical framework devoid of any consideration of the Judeo-Christian Creator. It may surprise you also that much of modern geology is not real science.

Many people either do not know the history of Western Civilization or they have forgotten it. What we are taught is that science has triumphed over religious superstition and in spite of religious persecution, has bravely investigated the geology of the Earth and has given us a true history of Earth. Is that really what happened?

Geologists use the Column, a man-made device, to help them make sense of the rocks and fossils buried all over the Earth. It is the standard by which every statement about geologic history, rocks and fossils is measured. Everything in geology today is determined by, fashioned and molded according to this *Bible* – from geology to archaeology, from the Big Bang to anthropology to radiometric dating. Having rejected the Biblical view of Earth history, another structure had to be developed to take its place.

Beginning in the early 1800s, as geologists drifted further and further from the Biblical view of Earth history, a uniformitarian view of rock formation, planet formation and biological formation began to take shape. Remember that uniformitarianism is the idea that present, observable geologic processes are sufficient to explain Earth history. Instead of a catastrophic explanation for the landforms around the Earth, an explanation that involved slow and gradual geologic processes was developed. The rock layers were viewed as past environments that once occupied the rock layers we now see - particular living things that were preserved as a record of past life and its evolution. As millions of years passed, some of the life was preserved as fossils. Hence the Column today is supposed to represent the record, however incomplete, of Earth history – geological and biological. Gone were the ideas of a six-day creation from Genesis 1, of the Fall in Genesis 3, and of the global Flood in Genesis 6-9. The entire foundation of the Bible and the message of a Savior were slowly eroded away, leaving a totally secular and Biblically contradictory story of the history of the Earth and living things.

But think about this – if God did not create us, then there is no accountability to Him. If the story of Adam's fall and sin were just myths, then gone is the need for a Savior. And if a Savior is not needed, then there is no heaven and no hell. Whether humans realize it or not, modern geology is an alternate philosophical explanation of man's existence and history versus the Biblical one. Is modern geology then science or philosophy? It is a competing philosophy; a worldview! It is for these reasons that we must examine The Geologic Column to see if, **(1)** it is a scientific truth, and **(2)** if it fits the facts of the rock layers.

The Development of the Geologic Column

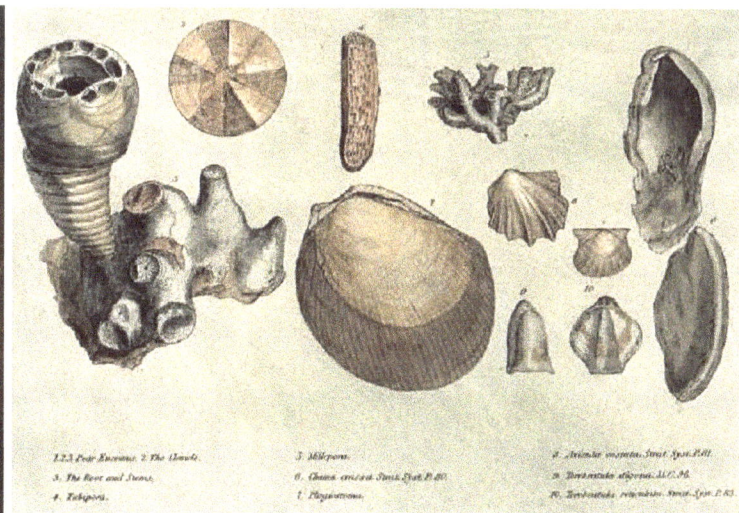

William Smith, the Father of English Stratigraphy is famous for drawing the first geological map of England. When it was first published it was overlooked by the scientific community because Smith was of relatively humble education and family connections. This prevented him from mixing easily in learned society. Consequently, his work was plagiarized and he was financially ruined. He even spent time in debtors' prison. It was only much later in his life that Smith received recognition for his accomplishments. The drawing on the right shows some of the fossils he catalogued from his work.

In 1815 William Smith, an engineer, was working on a canal in Britain. He took notice of the fossils he found in each of the stratum present. If Smith had been knowledgeable of the Genesis Flood, he would have interpreted these fossils as remnants of the global Flood. Instead, they seemed to him to have a pattern of fossil occurrences which he subsequently termed, The Principle of Faunal Succession. Faunal succession was one of the forerunners of evolutionary thinking.

Smith was a catastrophist. During his day this meant that he believed that the Earth was perhaps millions of years old and that it had been shaped by many global catastrophes in the past. During each period between catastrophes, God created new life. Smith thought that the layers of rock demonstrated this. What this story tells us, however, is that a particular ideological framework interpreted the evidence, not science. Without the Flood framework to interpret the rock layers, *any* framework could interpret the evidence.

As the Biblical explanation for Earth history was being eroded in the late 1700s, a new religious view of God arose called Deism – an impersonal Being who created, but is not involved in that creation. He had left the earth to run according to natural physical laws. This view was a sophisticated type of **atheism**. It was not a denial of the existence of God, but a denial of His continued involvement in His creation – practically though, it was atheism. It was a convenient and socially acceptable way of excluding God from the study of nature.

In this Deistic climate a new way of organizing the rock formations and fossils began to develop. In an age of increasing interest in nature, more and more time, effort and exploration took place in various parts of Europe. There was no geologic timetable or column to work with at that time. Individual geographical locations were studied by local geologists and names were given by those geologists for the rocks that were studied. The original locality names were retained for the most part. In the early 1800s the so-called Geologic Column looked like this:

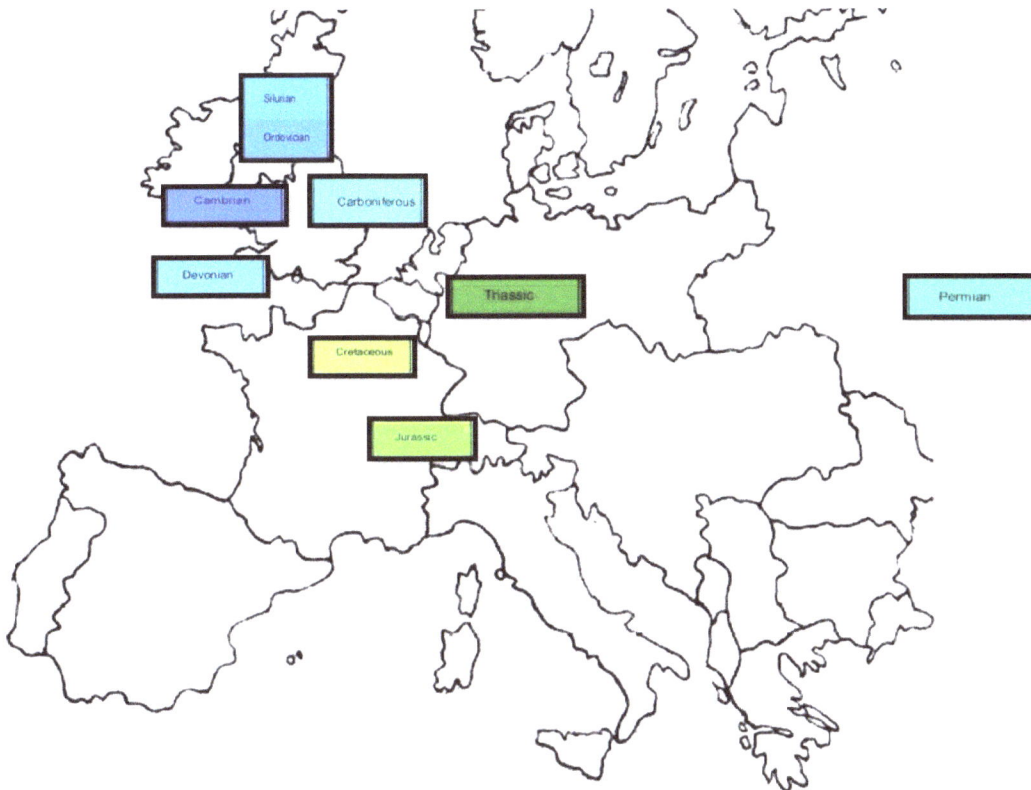

Essentially there was no column! All of the groups of rocks that were studied were not part of a series of rock layers; they were found all over Europe and Russia. Geology consisted of individual rock formations studied by independent geologists who knew their particular geographic area. There was no vertical column.

Observations about the original Geologic Column

- There was no particular order in studying the individual rock layers
- There was no initial order that presented a column
- The **Jurassic** was named after the limestone in the Jura Mountains in France
- The **Silurian** / **Ordovician** were named after after ancient Welsh tribes

- The **Cretaceous** was named from Latin, *creta*, meaning *chalk*, of the Paris Basin - 1822

- The **Cambrian** was the classical name for Wales
- The **Triassic** was named after the red beds, chalk, and black shales found throughout Germany and Northwest Europe - 1834
- The **Carboniferous** was named after the coal strata in England
- The **Devonian** was named after the English county of Devonshire
- The **Permian** was named after rocks in Perm, Russia
- There was no age or evolutionary significance inititally attached to the rocks in these various areas

The big question that challenged the early geologists and one that remains today is how should the rock layers be organized? This is called **correlation**. What do the rock layers mean? The early geologists set out to organized or correlate the various rock layers into an alternate view of Earth history that was in opposition to that of the Bible's. Correlation has been the goal of modern geology to this day!

Do these layers of rocks represent millions of years of time and past evolutionary life or do they represent one massive extinction event preserved in alternating layers of sedimentary rock, misinterpreted because of the uniformitarian rejection of the Flood? We will explore that as we go on in our study.

In the early 1800s it was discovered that different layers contained different fossils. Fossils were thought at that time to be representative remains of extinct life that had lived here for a time and then vanished. Early thinkers thought that the simplest forms of life were found in the Cambrian rocks. These fossils were sea creatures, and many had apparently become extinct. The early thinkers thought that these rocks must represent ancient life. These must be the oldest rocks. The Cambrian rocks became the Cambrian Period. The idea of a progressive column slowly began to dominate the study of the rock layers. The correlation of the various rock layers with their fossils became the goal in geology - line up the fossils from simplest to most complex and then an order to the rocks and fossils would tell the story of Earth history.

It was most likely a little more complicated than that, but the belief in **fossil succession** is what ultimately decided the order of the various rock layers. There were no dates imprinted on the layers of rock! The ages for the various rock layers began to be assigned when it became apparent to these early geologists that life must have taken millions of years to develop from simple to complex. The Column has been based completely on the assumption that life arose through hundreds of millions of years of evolutionary development.

Another important factor in assigning ages to the Column came after radioactivity was discovered in 1896. In 1907 the first use of uranium to date rocks resulted in ages of 400 million to 2.2 billion years old. Look at the following diagram of the "evolution" of The Geologic Time Column.

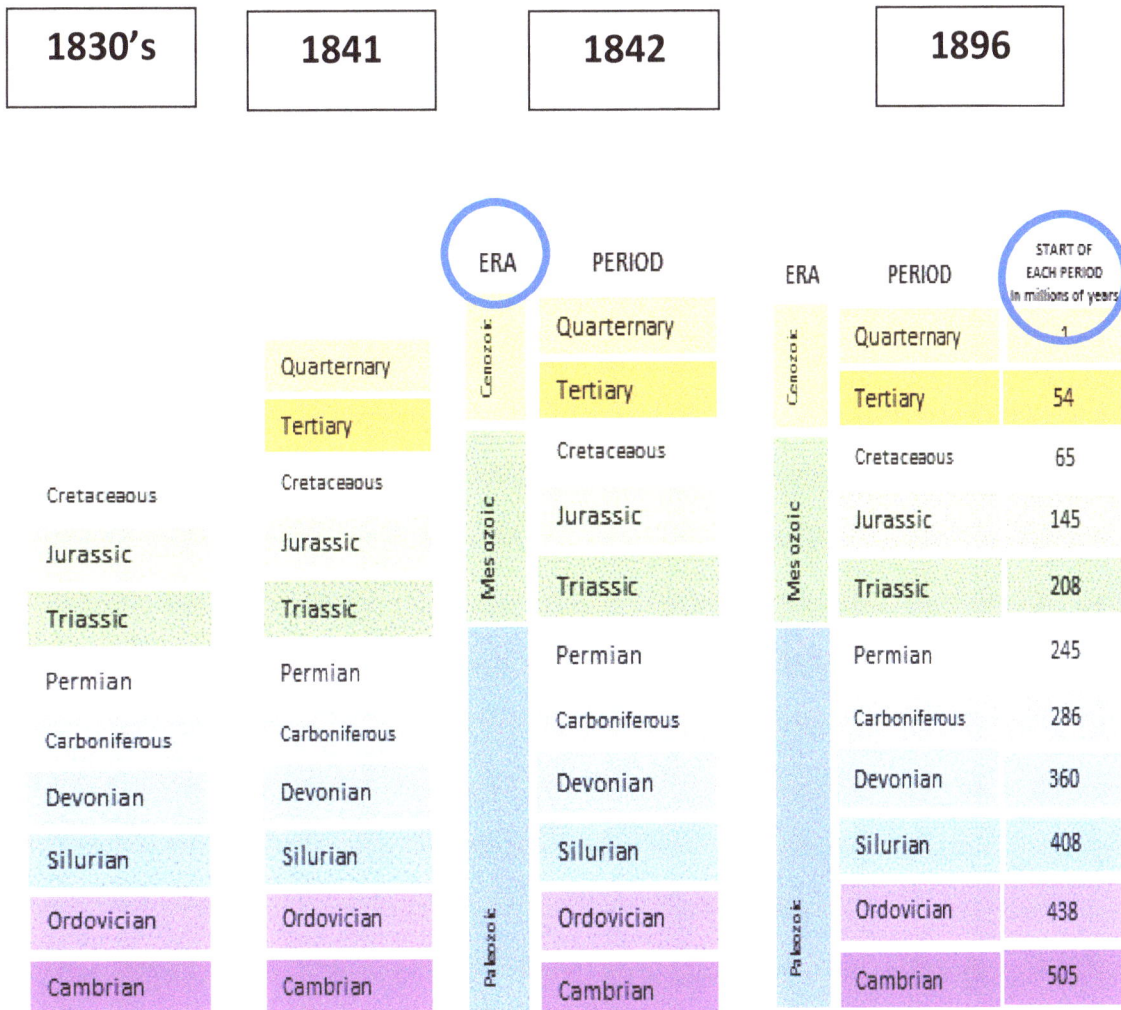

The "Evolution" of The Geologic Time Column

| 1830's | 1841 | 1842 | 1896 |

ERA	PERIOD		ERA	PERIOD	START OF EACH PERIOD in millions of years
Cenozoic	Quarternary		Cenozoic	Quarternary	1
	Tertiary			Tertiary	54
Mesozoic	Cretaceaous		Mesozoic	Cretaceaous	65
	Jurassic			Jurassic	145
	Triassic			Triassic	208
Paleozoic	Permian		Paleozoic	Permian	245
	Carboniferous			Carboniferous	286
	Devonian			Devonian	360
	Silurian			Silurian	408
	Ordovician			Ordovician	438
	Cambrian			Cambrian	505

1830's column: Cretaceaous, Jurassic, Triassic, Permian, Carboniferous, Devonian, Silurian, Ordovician, Cambrian

1841 column: Quarternary, Tertiary, Cretaceaous, Jurassic, Triassic, Permian, Carboniferous, Devonian, Silurian, Ordovician, Cambrian

The "Evolution" of The Column – explanation

- **The 1830s** – the Geologic Column is organized according to the assumed history of life succession. The fossils were arranged according to the assumption that life arose over time and then changed. This scenario seemed to be able to explain the reason for "less complex" life on the bottom of rock layers and "more complex" life on the top.
- **In 1841** Charles Lyell proposed adding another couple of Periods that would incorporate the sediments and fossils of an apparent ice age.
- **By 1842** Charles Lyell had added additional time periods to the Column that he called Eras. The word era describes a vast amount of time and also reflects the assumed life succession now being assigned to the fossils. Paleozoic = *ancient life*, referring to all of life in the sea as being the first and the oldest life; Mesozoic = *middle life*, also called, *The Age of Reptiles*; Mesozoic probably referred to the transition from sea life to land life or between the sea creatures and the mammals; and, Cenozoic = *new life* or *recent life*, referred to as, The *Age of Mammals*. In 1859 Darwin published his *On the Origin of Species* which had the effect of uniting geology and the biology into a complete worldview.
- **In 1896** radioactivity was discovered giving a supposed means of dating the ancient past. Using the decay of uranium, the age of the Earth was set at between 400 million and 2.2 billion years in the early 20th Century. This is a far cry from the modern declaration that the Earth is 4.6 billion years old!

What does the Column really tell us?

The Geologic Column is an attempt by modern geologists to organize the various rock layers and the fossils in them. It portrays an upward evolutionary story for the history of life and of the Earth. The pictures of the plants and animals that often accompany the Column imply that this is exactly how Earth history happened, that different life forms evolved during the various geologic periods listed and that is the end of the debate. But the Column is really nothing more than an alternative, competing story to that of the Bible. It is a very sophisticated creation of man.

As the Geologic Column is based on the fossils found in different layers, we will now focus on that fossil evidence.   Wait to take the quiz until you are done with Lesson 6.

Lesson 6 – The Geologic Column and the Fossil Evidence

The aged Charles Lyell – the most influential man in the history of the development of modern geology. He was determined to eradicate any influence from Genesis in the development of modern geology.

Words to know: *taxonomy   speculation*

The Geologic Column is in reality the development of an idea. The portrayal of The Column is mostly **speculation**. Very rarely will geologists find any of the layers stacked in what they see as the correct order. The layers are correlated or put together according to an assumption that biological evolution is a true scientific fact. In fact, there are all kinds of exceptions to this ideal. Here are a few:

1. **Out of place fossils** – fossils showing up where they should not be according to modern evolutionary thought. Often the Geologic Column is portrayed with the plants and animals pictured to the right of the various Eras and Periods. A picture is worth a thousand words! This does give the impression that this arrangement is exactly how geologists find the various fossils, all stacked in a progressive arrangement as the Theory of Evolution teaches. But there are many out of place fossils.

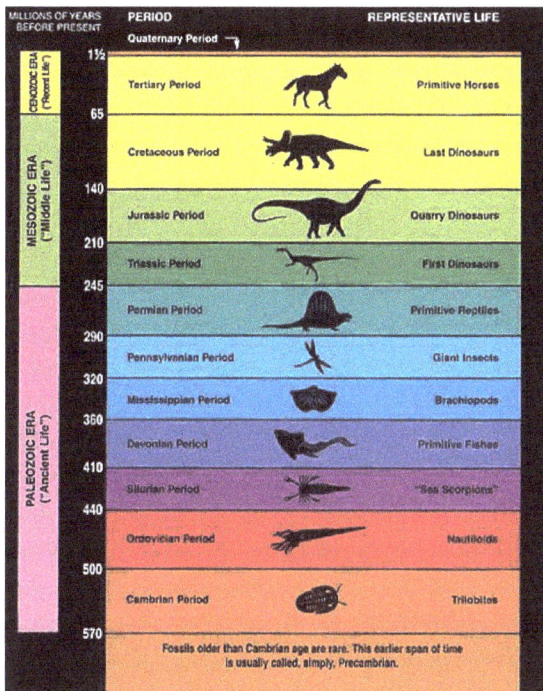

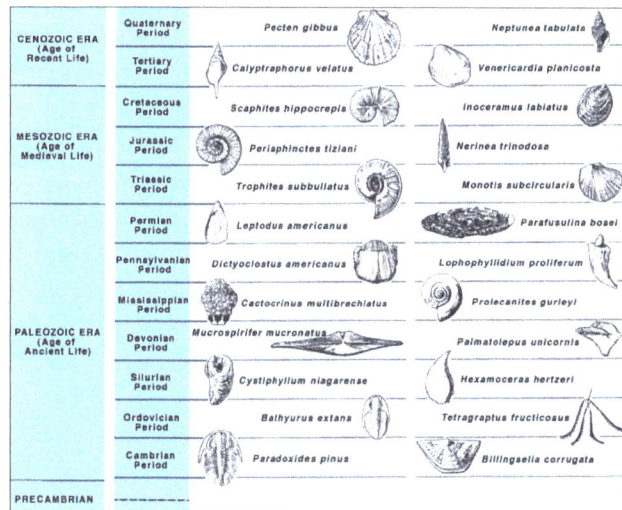

The Geologic Column as it is often presented: the one on the right uses Index fossils, which we will discuss later.

- The Wollemi Pine – thought to have gone extinct more than 150 million years ago during the Jurassic Period on the Column. A few years ago it was found alive and well growing in Australia. This indeed surprised geologists because the fossil record is supposed to have recorded life that once existed and then either drastically changed into something else or went extinct. One researcher commented on the discovery of the live Wollemi Pine that, "It was like finding a live dinosaur." Would this not be better explained by the Genesis Flood as catastrophic burial during Noah's Flood about 4,500 years ago as well as survival and regrowth since then?

Wollemi Pine

2. **Wrong order** – younger fossils found below older.
  - For many years it was thought that large mammals only evolved after the dinosaurs began to die out. However, a modern find of a fossil mammal contained a small dinosaur in its stomach! This discovery not only puts the evolutionary development of larger mammals much earlier than thought, but seems to dispel the notion that only small mammals existed during the time of the dinosaurs.

Dinosaur Fossil Found in Mammal's Stomach. This find has been dated at 130 million years ago by secular geologists. This is significant because it shows that mammals capable of eating dinosaurs were around during the Cretaceous, the time of Tyrannosaurs Rex!

3. Mixed fossils – animals in habitats where they should not be. These have been particularly true of bone beds where a number of different fossils from different animals had been buried or collected together. This has been true of bone caves where it appears that bones from different animals had been washed into the cave together.

- One of the best examples has been the Cumberland Bone Cave in Maryland, originally discovered in 1912.

A paleontologist shows a tray of varied vertebrate and invertebrate fossils from a variety of animals all jumbled together into the Cumberland Bone Cave, indicating a catastrophic event of some kind.

Many of these fossils are of varieties that do not live in the same vicinity of one another today.

- Another example is the Ginkgo Petrified Forest of Washington State. In fact, this is not a forest. Over 200 different kinds of wood are meshed together in the same location. Most of these woods grow in different environments today. Would this not be better explained by a watery catastrophe that had washed in trees from various environments?

Petrified spruce log, Ginkgo Petrified Forest.

4. Living fossils – plants and animals thought to be extinct for millions of years, only to show up alive, well and looking every bit the same as their supposed fossil ancestor.

- The classic example of this is the coelacanth fish which supposedly died out with the dinosaurs 65 million years ago. However, in the 1930s Japanese fishermen began catching it off the coast of Madagascar. The fish was exactly like its fossil predecessor; there had been no evolutionary change in 65 million years!

A living coelacanth and its fossil on the right.

- Velvet worm a living fossil – The velvet worm is dated, according to modern evolutionary thought, to early Cambrian, over 500 million years old, yet exists today virtually unchanged!

Velvet worm

5. Extension of fossil ranges up and down the Column – A fossil range is that time frame over which a creature is thought to have lived. Fossils of creatures that appeared to have lived for a while and then went extinct were thought to have limited time frames. This is portrayed in The Column, generally to the right of the Eras, Periods and Epochs. This was called the fossil's range.

- A type of theropod dinosaur called *Alvarezsaurus* – recently a dinosaur was discovered in Argentina and was dated, according to conventional dating methods, at 63 million years earlier than it was supposed to have existed. This is significant because it is thought to have been in the line leading to birds, throwing off the whole evolutionary scheme of dino-to-bird evolution.

Note this quote about Alvarezsaurus from an evolutionary article: "Haplocheirus is a genus of alvarezsauroid theropod dinosaur - It is the most basal member of its clade. It is the oldest known alvarezsauroid, predating all other members by about 63 million years. This also makes it about 15 million years older than the oldest known bird Archaeopteryx."[7] Even they acknowledge that there is a problem in the dating!

With more collecting since the 1800s an increasing number of the same fossils are showing up in other Periods, giving the appearance that they went extinct for a while and then re-evolved again and thus found again in an entirely new Period.

- Examples include shark fossils - they have been discovered in rock layers that have supposedly been dated 15 million years before sharks were supposed to have evolved. In 2003 it was reported by National Geographic that the world's oldest shark had been discovered in 409-million-year-old rock layers in New Brunswick, Canada. The fossil found is called Doliodus problematicus. Why would they name it problematicus? Is it because this pushes the beginnings of shark evolution even further back? Certainly, a long evolutionary history is implied. Since all forms of life share a common ancestry, what might this do to other forms of life that supposedly evolved into or from sharks? The details of the whole story have to be changed

[7]Web. 15 Jan. 2015. ‹http://en.wikipedia.org/wiki/Haplocheirus›.

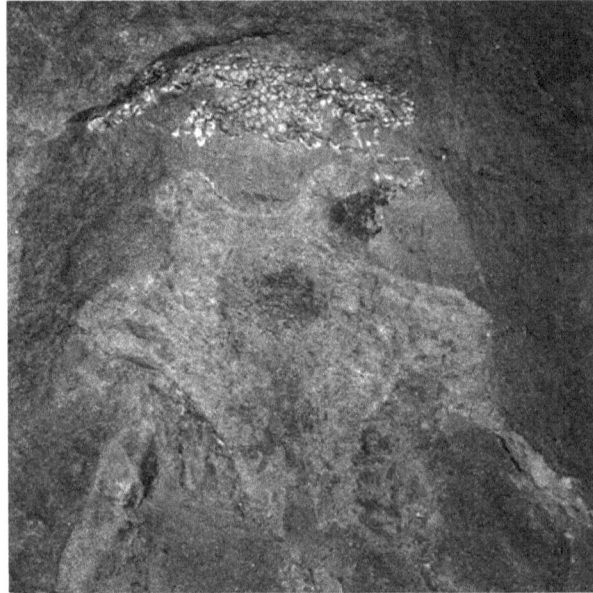

Doliodus problematicus

- In 1996 scales from the jawless fish Anatolepis were found in the Cambrian along with trilobites. Fish were not supposed to have evolved until the Devonian, 150 million years later. How would this affect the creatures that supposedly evolved into or from Anatolepis? Would this not throw off other aspects of The Column?

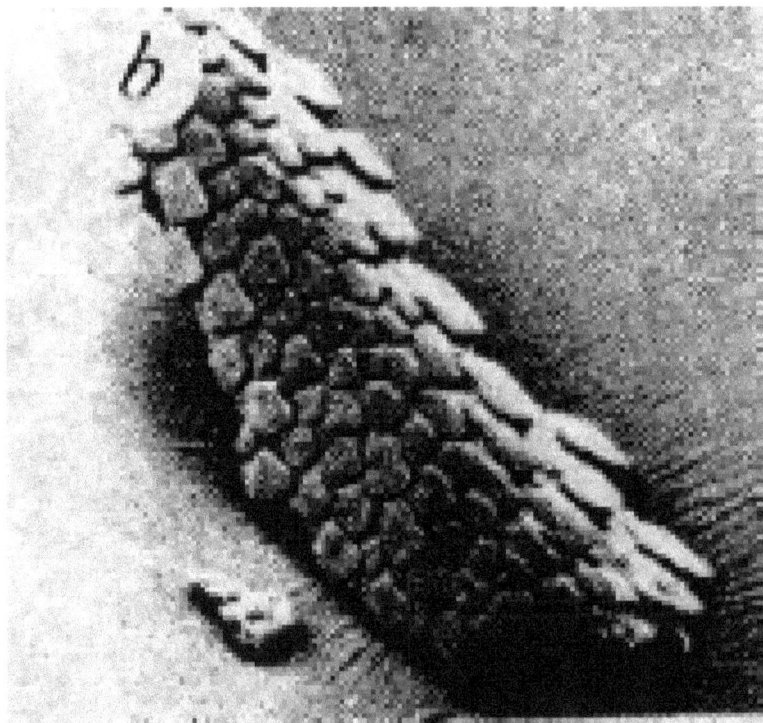

Anatolepis

- Another case involves one of the major icons of fish to amphibian evolution – Tiktaalik (pronounced, *tick–TAH-lick*).

Tiktaalik, the fossil – In the April 6, 2006 issue of <u>nytimes.com</u>, an article appeared entitled, "Fossil Called Missing Link From Sea to Land Animals." And so Tiktaalik was immortalized as a genuine missing link between fish and amphibians.

Evolutionists were convinced that fish gave rise to amphibians long before Taktaalik was discovered. Paleontologists had been looking for just such a fossil. The April 6, 2006 article in <u>nytimes.com</u> says:

> Scientists have discovered fossils of a 375-million-year-old fish, a large scaly creature not seen before, that they say is a long-sought missing link in the evolution of some fishes from water to a life walking on four limbs on land. [8]

The article goes on to describe it as a large *fish* but with features that paleontologists are convinced show transition from fish to amphibian. All the fossils found of Tiktaalik are just like the one above. I think that before we call this creature a transitional link, we would have to see a few more stages of evolution preserved in the fossils. Another way of looking at this creature is that he was all fish – an extinct and strange creature, but all fish.

The two images below illustrate the supposed transition between Ichthyostega, (pronounced *ik-thee-oh-STAY-guh)*, and Tiktaalik. It looks very compelling. In reality it appears that we are looking at three different creatures whose relationship to one another has not been proven except in the imagination.

---

[8] Wilford, John. "Fossil Called Missing Link From Sea to Land Animals." 6 Apr. 2006. Web. 13 Jan. 2015. <http://www.nytimes.com/2006/04/06/science/06fossil.html?pagewanted=all&_r=0.>.

Late Devonian lobe-finned fish and amphibious tetrapods.

land

Tiktaalik

Ichthyostega

rivers,
swamps and
shallows

Panderichthys

Acanthostega

Eusthenopteron

Coelacanth

sea   millions of years ago
385   380   375   365   360

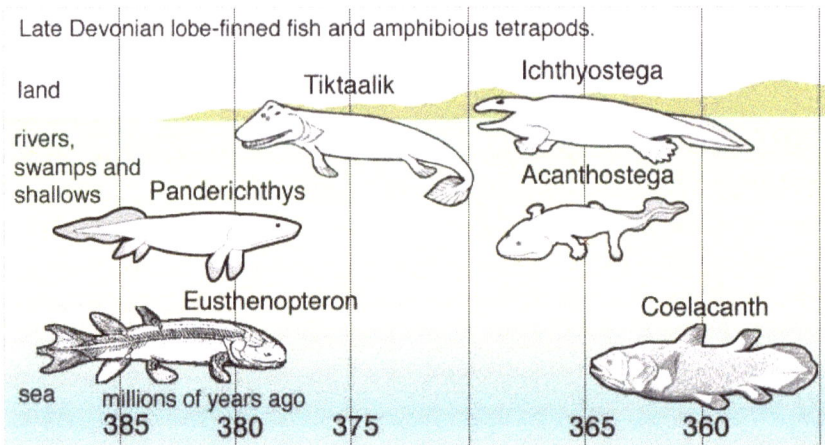

From Fish to Amphibian?

Eusthenopteron Fish

Tiktaalik
Transitional link
or extinct fish?

Ichthyostega
"Fish roof"
Extinct fish
or 4-legged animal?

Now, here is the startling news! In 2010 an article appeared with some upsetting news.[9] The trackway of a four-legged land animal was found in rocks that have been dated at 397 million years old.   What is so strange about this? Tiktaalik has been dated at 375 million years old. The fossil trackways of the four-legged land creature are 22 million years older than Tiktaalik! Simply put, Tiktaalik cannot be the transitional link between fish and four legged amphibians! This find would push the evolutionary history of land-dwelling four-legged amphibians back even further!

---

[9] Bryne, J. "Four-legged Creature's Footprints Force Evolution-rethink." *LiveScience.com.* 6 Jan. 2010. Web. 9 Jan. 2015. ‹http://www.livescience.com/6004-legged-creature-footprints-force-evolution-rethink.html›.

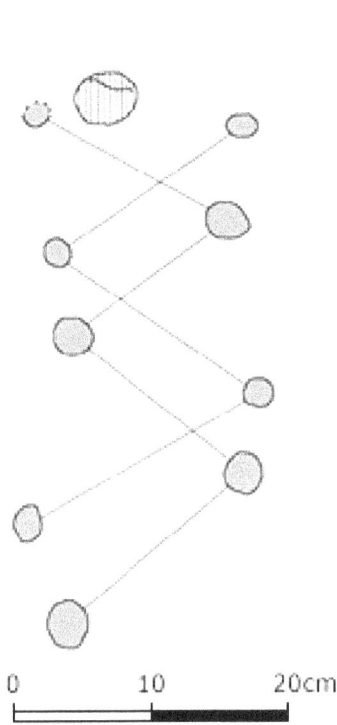

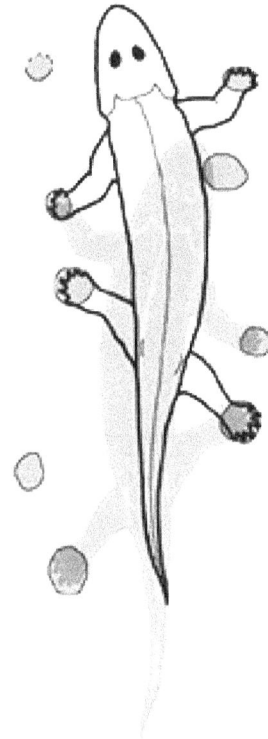

397 million year old trackway of a land-dwelling, four legged creature that came 22 million years before Tiktaalik.

An alternative explanation for Tiktaalik is that he lived during pre-Flood times in a transitional zone unlike any that exists today, a zone of thick, underwater vegetation. One paleontologist comments that it seems to have had pseudo-hind limbs that enabled it to navigate difficult waters. It may have had swim bladders like today's lungfish, but it had scales and lived in the water. It was a fish, not a "fishibian."

- Fossil bee nests found in the Late Triassic – found in petrified trees from Petrified Forest National Park in Arizona dated at 220 million years! The earliest known bee nests up to that point were found in rock layers dated around 96 million years old. Flowering plants, which bees require, were not supposed to have evolved until 100 million years later. What does this say about the supposed evolution of bees? This discovery is still being investigated.

A

B

Fossil bees' nests found in rock layers dated 124 million years before bees were supposed to have evolved. Flowering plants, which bees need, were not supposed to have evolved for another 100 million years. How can this not affect the whole picture of evolutionary relationship? ("A" and "B" in the picture are from the original image.)

- 320-million-year-old amber discovered! Amber is mainly found in strata classified as Cretaceous and Tertiary (around 100 million to 40 million years old). But just recently amber was found in Carboniferous coal in Illinois and was dated 320 million years old within the uniformitarian timescale. Such a date is one of the oldest, if not the oldest, for amber. The Carboniferous is supposed to be the time that many plants, now extinct (such as lycopods), ruled the swamps and forests. The chemistry of the amber turned out to be that observed only from flowering plants (angiosperms) that supposedly had not yet evolved. It would take another 200 million years for angiosperms to evolve!

Many modern-looking insects are wonderfully preserved in amber, making it an ideal record of life at the time the amber was formed.

6. Stasis – stasis means holding steady; no change. The fossil record is filled with examples of creatures that have been dated to be millions of years old and yet their living counterparts show no evolution at all. It is as if they were frozen in time and stopped evolving.

A fossil magnolia leaf and its live counterpart; fossil dated in millions of years.

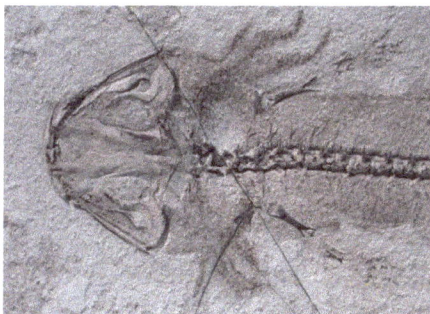

A fossil salamander dated at 160 million years old, and a live salamander.

A fossil shark dated at 400 million years old, and a live shark. It shows no evolutionary change in all that time. The next picture shows a fossil of a snake supposedly dated to be millions of years old and a living snake. These examples show stasis or stagnate evolution.

Scorpion Fly 164 million years old still recognizable and classifiable – no discernable evolutionary change in all that time!

Fossil cricket from the Cretaceous (over 65 million years old) shows no change from a modern cricket.

7. Fossil skipping. What is fossil skipping? A creature that is found as a fossil in one geological period of rock layers disappears. Its absence continues throughout other periods of rock layers above its discovery, then, reappears in later periods of rock layers without any change.

- A few land snails are found in the Carboniferous (about 350 million years ago). No land snails have been found in the Permian, Triassic or Jurassic above it. They then reappear in the Cretaceous, over 205 million years later and then are continuous to the present.
- A few scorpions are found in the Upper Silurian (about 410 million years ago). None have been recognized from the Devonian, immediately above it. But in the Carboniferous, about 50 million years later, both scorpions and spiders occur. Both these groups appear to be missing from the Permian and from the whole series of the Mesozoic Era, spanning about 150 million years. They reappear in the Tertiary, directly above the Cretaceous.
- Insects have been around for about 300 million years. But during the Cretaceous (a period of about 73 million years) their fossil record is all but absent.

8. Taxonomic manipulations – *taxonomy* (from Ancient Greek: taxis, *arrangement*, and -nomia, *method* or name) is the science of defining groups of biological organisms on the basis of shared characteristics and giving names to those groups. The modern taxonomic system was originally developed by Carl Linnaeus in the 1700s. It is not necessarily the way things really are, but is an attempt to classify living things. In Genesis chapter one God gave us taxonomy, or *kinds*. This was rejected during The Enlightenment and remains rejected today. Through the years, attempts have been made to classify all living things, including those things which are extinct. Now, it should be apparent that to classify living things which can be observed is one thing. We call these things *extant*, meaning *to stand*, from the Latin, *ex stare, to stand out*. To classify extinct things where we cannot observe habits and reproduction is another. Modern taxonomists are left to guess based on subjective opinion. This has been the case in the classification of extinct things such as dinosaurs. As we shall see, it is extremely difficult to apply the modern concept of taxonomy to fossils.

Modern taxonomy assumes that all things have evolved from a common ancestor and so share similar characteristics. These similarities are supposed to demonstrate relationship. Evolution assumes that through the eons of time, living things have changed from one group to another and so share similar characteristics. But again, this is all subjective. It really depends on those who have the authority for deciding what is and what is not related.

- **Cephalopods**, meaning *head-footed*, have been the subject of an enormous amount of debate. The tiniest variation in the shell of one of these creatures has been used to reconstruct its taxonomic history. This has been proven to be premature as more and more fossils have been discovered that cross once hallowed taxonomic lines.

Michael Oard says, "Minute changes in cephalopods have been used to date Mesozoic strata. But there is evidence that many defined species are time-transgressive (i.e. varying in age in different areas, or cutting across time planes), so cannot be used to define a certain, exact time period in the geological timescale. Undoubtedly, the defined species are subject to taxonomic manipulations, since it is very difficult to define a species today and impossible with just bones or shells since hybridization tests cannot be performed."[10]

- **Protozoa** - Another example of taxonomic manipulation is the foraminifera or protozoa. Fossil protozoa are prolific. There are many variations of protozoa which have been used to date certain strata. However, like most fossil collecting, more and more exceptions are being found all the time.

    Within a single species the foraminifers may have thick ornamented walls under normal oxygen concentrations, and thin, less-ornamented walls under low oxygen conditions. ... Because of the many examples of variation in living and fossil forms, foraminifers are considered to be extraordinarily plastic. A foraminifer may contain enough genetic information to express many different forms, depending on the conditions. [11]

---

[10] Oard, Michael. "Taxonomic Manipulations Likely Common." *Journal of Creation* 25(3): 15-17 December 2011

[11] Tosk, T. "Foraminifers in the Fossil Record: Implications for an Ecological Zonation Model." *Origins* 1 Jan. 1988: 8-18. Print.

Because of the amount of variation that is being discovered, the question in the case of the protozoa is, what time Period should they be assigned? Plasticity implies a great degree of adaptability and variation. What criteria are going to be used to date these animals and arrange them in an evolutionary progression? How much subjectivity will be involved in placing the protozoa into evolutionary niches?

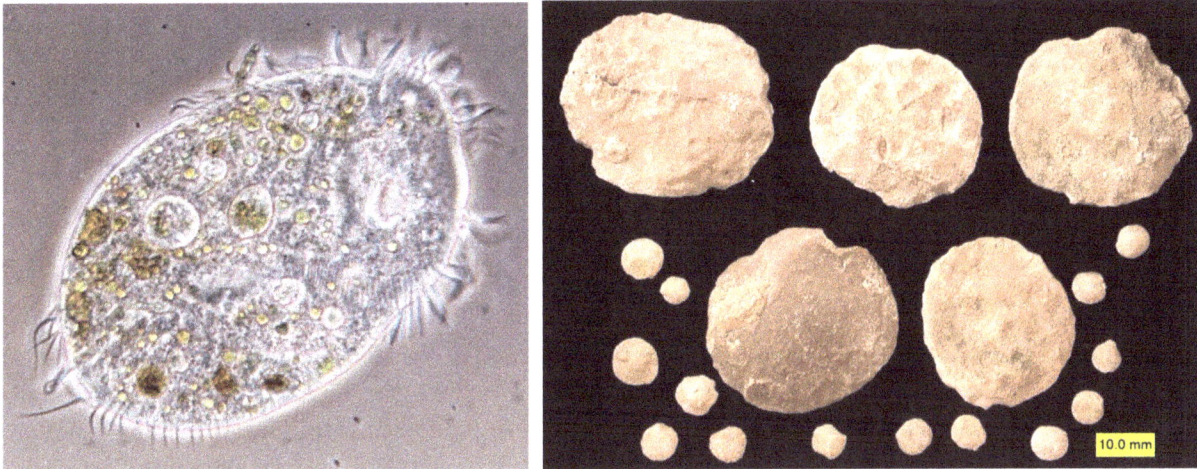

Protozoa, live and fossil

Index fossils

What are index fossils? Scientists have used *index fossils* to measure time boundaries for The Geologic Column. Index fossils are those fossils which are used to age-date strata of different compositions, for example shale and limestone. How do geologists do this? The idea of using index fossils as age indicators was developed long before Darwin. Geologists noticed that certain fossil marine creatures occupied certain strata and then disappeared or appeared differently in other strata. Without consulting the mechanism of the Genesis Flood as a geological explanation, it was assumed that certain creatures lived for a period of time and then became extinct, having evolved into something else. So, the assumption of fossil succession drove the use of fossils to age date the strata. Geologists merely assumed that the strata that contained the same fossils were of the same age, regardless of the composition of the strata. This branch of geology has been called biostratigraphy which focuses on correlating and assigning relative ages of rock strata by using the fossil assemblages contained within them. The aim is correlation of the different strata.

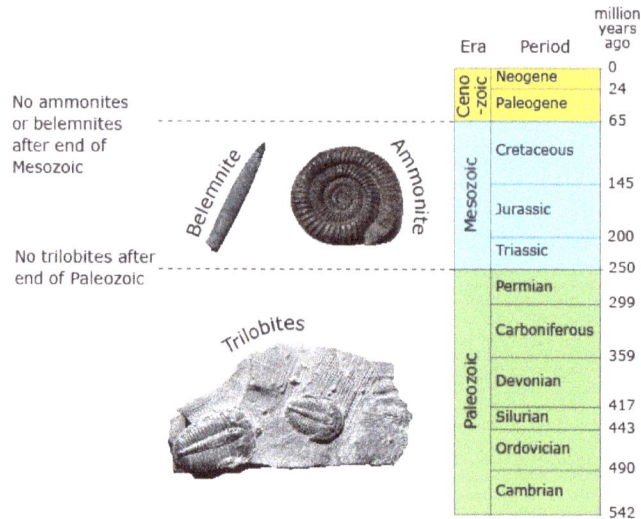

Era | Period | million years ago

Cenozoic: Neogene (0, 24), Paleogene (65)

Mesozoic: Cretaceous (145), Jurassic (200), Triassic (250)

Paleozoic: Permian (299), Carboniferous (359), Devonian (417), Silurian (443), Ordovician (490), Cambrian (542)

No ammonites or belemnites after end of Mesozoic

No trilobites after end of Paleozoic

Belemnite   Ammonite

Trilobites

This illustration from The University of Wisconsin emphatically states that certain animals or plants exist for a while and then die out. This is a rule of fossil succession. It is believed, therefore, that their fossil remains can be used to correlate the different strata into time niches. But how hard and fast is this? As we have seen, many fossils have been time transgressive. Are these index fossils time indicators or ecological niche indicators as the result of the general order of Flood burial?

A legitimate question to ask is, could the rock layers actually be telling a different history – a story of global catastrophe that took place in Earth history a short time ago and wiped out masses of plants and animals, churned them up, transported them, buried them in the Flooding Stage of the Flood, and then tore them up again during the Receding Stage of the Genesis Flood? Only a prior commitment to a uniformitarian set of assumptions will preclude any such interpretation.

Dinosaur National Monument, UT
Morrison Formation, Late Jurassic

A catastrophic watery event best explains the massive and numerous fossil bone beds all around the world. (L) The Fossil Hills' waterhole bone bed as depicted in the visitor center and museum's diorama exhibit at the Agate Fossil Beds National Monument. (R) Dinosaur National Monument.

Fossil bone bed, Bighorn, Wyoming.

One of the greatest indicators of a watery catastrophe is the presence of whole fossil trees buried through successive layers of strata. These trees have been called *polystrate,* meaning *many layers.* The layers or strata in which they are found are supposed to represent different geologic ages. How can that be? Would it not make more sense to interpret these fossil trees as having been buried as successive layers of muddy sediments were laid down in rapid succession?

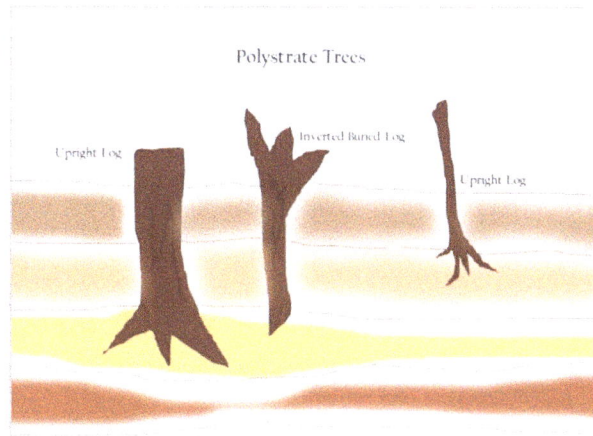

One of the greatest anomalies in naturalistic uniformitarian geology is the presence of polystrate trees all over the world. These are fossil trees that have been preserved through layers which are supposed to represent time markers. How could trees last for millions of years while environments were preserved in rock layers as they accumulated across time? They couldn't. They would undergo decay and quickly erode away, leaving only a remnant from a particular time in the geologic past. Trees preserved in successive layers indicated rapid deposition of layers of mud and watery sediments which covered trees torn up by the Genesis Flood.

As challenges to The Geologic Time Table continue to mount, let's take another look at geologic time as interpreted by the framework of the Genesis Flood. Below is a representative geologic column from a Biblical perspective. A cursory reading of

Genesis 1-9 will give us a framework from which to interpret the various land forms all around the world.

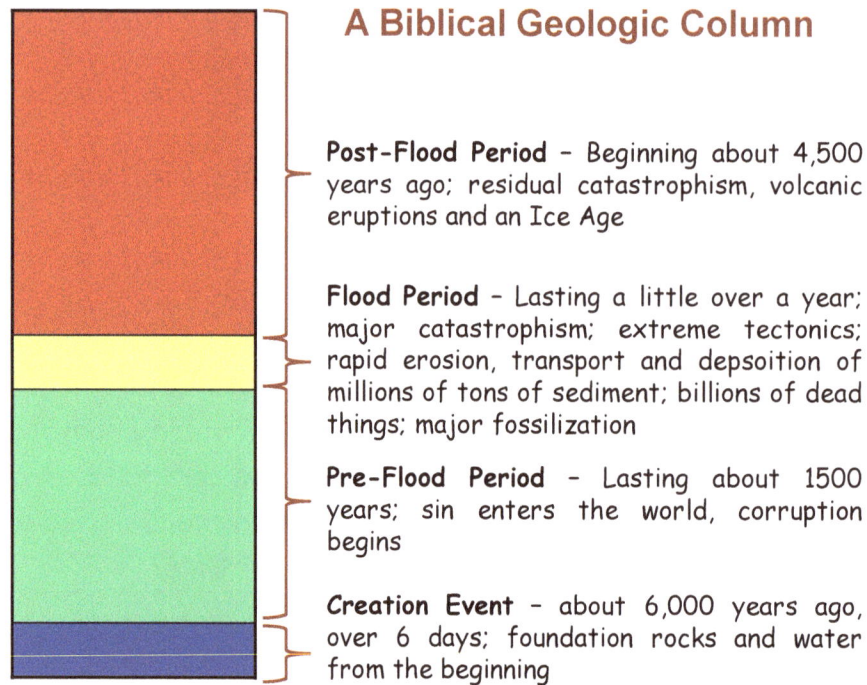

A Biblical Geologic Column

Post-Flood Period – Beginning about 4,500 years ago; residual catastrophism, volcanic eruptions and an Ice Age

Flood Period – Lasting a little over a year; major catastrophism; extreme tectonics; rapid erosion, transport and depsoition of millions of tons of sediment; billions of dead things; major fossilization

Pre-Flood Period – Lasting about 1500 years; sin enters the world, corruption begins

Creation Event – about 6,000 years ago, over 6 days; foundation rocks and water from the beginning

Activity 6

Using internet resources make a list of the index fossils geologists use today as time markers. For each fossil, write out a Genesis Flood alternative explanation for its placement in the Geologic Column.

Please take Quiz #5 and #6, Appendix B

Lesson 7 – Dinosaurs and Geologic Time

Words to know: *dinosaur   asteroid   disarticulated*

Likely major causes of mass extinction:

Flood basalt events (volcano eruptions)

Asteroid collisions

Sea level falls

Other possible causes of mass extinction:

Global warming

Global cooling

Methane eruptions

Anoxic events

According to National Geographic, scientists have narrowed down the most likely causes of mass extinction.[12] Notice that the Genesis Flood is conspicuously missing. It is interesting to note that all of the above causes of extinction ("global warming" and "global cooling" being seen as climate change) could be indirect causes of the one main cause – The Flood.

Today we are dogmatically told by scientists that the **dinosaurs** died out 65 million years ago due to a large **asteroid** smashing into the earth resulting in the death of the dinosaurs along with some animals too. Now, if this is true, then the Bible cannot be true. So, we need to address this. We have already concluded that the Bible is historically accurate and reliable. The error must be in the way scientists view earth history and dinosaurs.

One of the main objections that some paleontologists have made against the asteroid idea is the selective nature of the extinction event. One college geology textbook describes this extinction event this way:

---

[12] *Mass Extinctions*. No author cited. National Geographic, found at https://www.nationalgeographic.com/science/prehistoric-world/mass-extinction/

Geologists speculate that some mass-extinction events reflect a catastrophic change in the planet's climate, brought about by unusually voluminous volcanic eruptions or by the impact of a comet or asteroid with the Earth. Without the warmth of the Sun (sic), winter-like or night-like conditions would last for weeks to years, long enough to disrupt the food chain. Either event, in addition, could eject aerosols that would turn into global acid rain, scatter hot debris that would ignite forest fires, or give off chemicals that, when dissolved in the ocean, would make the ocean either toxic or so nutritious that oxygen consuming algae could thrive.[13]

In view of the devastating consequences of this kind of event, how could it be that only the dinosaurs were the major category of animal affected by this catastrophe? Immediately into the next Era, the Cenozoic Era, called The Age of Mammals, evolution suddenly exploded producing a huge diversity of mammals and plant life. Meanwhile many plants and animals survived this horrific event and seemed to have thrived because of it. All this stretches credulity. Certainly an event of this magnitude would have done a lot more damage!

An asteroid slamming into the Earth like the artist's rendition above, would certainly do some damage. But what would have happened to the rest of the creatures living at the same time. Why was it selective? Why did mammals not only survive but thrive and diversify? Some paleontologists have pointed this out and have begun to look for other causes for the disappearance of the dinosaurs!

We must take a fresh look at dinosaurs and see if the Bible may have a better explanation for their disappearance.

History of the word dinosaur

The word dinosaur comes from the Greek words, *dino* – meaning *powerful* or *terrible* and *saurus* - meaning *lizard; terrible lizard*. It was invented by a man named Sir Richard Owen in 1842. Richard Owen, (1804 – 1892) was an English biologist,

---

[13] Marshak, Stephen, *Essentials of Geology*, Third Edition (W. W. Norton and Company, 2009), p. 510.

comparative anatomist and paleontologist. He opposed Darwin's evolutionary ideas and was a convinced creationist. There are historical records of living dinosaurs having actually been seen by man, and their fossils were definitely known before 1842.

Before the word dinosaur was coined, they were called *dragons*, from the Hebrew word *tannin*, meaning beast or dragon. And thus it is translated in many places in the King James Version of the Bible, starting in 1604 and completed in 1611. The King James Bible has been the standard and most popular translation of the Bible since then. This has had a huge impact on what people have called these strange and large beasts.

Little was known about dinosaurs at this time and the word *dinosauria* meant *terrible lizards*. Today the word has been reinterpreted to mean *terrible reptile*, because scientists believe dinosaurs were not lizards but reptiles. Of course, we don't really know what dinosaurs were or even how to accurately classify them, since they are not around or accessible to scientifically test their breeding or living habits. One historical thing seems certain – many people saw and recorded the existence of these beasts throughout human history. There is no scientific way to prove that dinosaurs died out 65 million years ago.

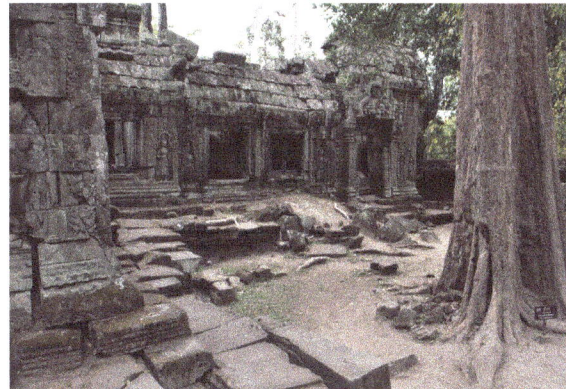

Many pictures of animals are recorded at one of the largest temple complexes in the world, Angkor Wat, built in the 12th century AD in (modern day) Cambodia. There is a picture of a live Stegosaurus carved into one of these temples at Ta Prohm.

An amazing discovery was made in 1925 by the German archeologist Waldenar Julsrud, when he discovered numerous dinosaur figurines buried at the foot of El Toro Mountain, in the Valley of Acambaro, about 180 miles North of Mexico City.

This was positive proof that the ancient civilizations of people, who lived in these areas several thousand years ago, had intimate first-hand knowledge of what these dinosaurs looked like. These dinosaur figurines are authentic proof that these monsters still inhabited the jungles of South America during the times of the Inca's. There are all different sizes of ancient dinosaur figures in this collection, some as big as five feet long.   Several hundred different dinosaurs including duck billed, Gorgosaurus (a type of tyrannosaurus), horned Monoclonius (a type of ceratopsian), Ornitholestes (a small theropod dinosaur), Titanosaurus (a type of sauropod dinosaur), Triceratops, Stegosaurus paleocococincus, Diplodocus, Podokosaurus (most likely a type of theropod dinosaur), Struthiomimus (similar to a theropod dinosaur), Plesiosaur, Leviathan, Maiasaura (a type of hadrosaur duck-billed dinosaur), Rhamphorynchus (a type of pteranodon), Iguanodon, Brachiosaurus, Pteranodon, Dimetrodon, Ichthyornis (an extinct bird), Tyrannosaurus Rex, Rhynococephalia (a lizard-like animal) and other unknown or yet unidentified Dinosaur species.   Over 33,500 figurines and artifacts were found including a collection of almost every known dinosaur found to date.

This find is considered controversial by evolutionists, primarily because it does not fit their geologic time scale. If this was the only example of dinosaur sightings by man, then there might be some question. But there are many such examples.

Clay figurines of various dinosaurs from 13ᵗʰ Century Mexico.

The major question of course is how people could have known what dinosaurs looked like in ancient times unless they had seen them. In a few places someone might have guessed based on bones they had found, but entire skeletons would have been extremely rare. Most fossil dinosaur bones are found in a *disarticulated* state. Books have been written about various dinosaur sightings, but they are dismissed by paleontologists today because they contradict the currently held view that dinosaurs went extinct 65 million years ago. The vast amount of evidence indicates otherwise.

What about the accounts of fire-breathing dragons? Certainly, those are false and just myths, aren't they? It is true that the imagination of man could conjure up beasts that might be similar to dinosaurs. But what would they be based on? In almost every myth there is a historical foundation. Could it be that people actually saw these creatures and then tried to preserve the memories of them through carvings and drawings? Most people are familiar with the story of St. George the dragon slayer. It is a popular icon of Christian history.

Left to right: Saint George the Dragon Slayer, early 16th century Russian icon; St. George and the Dragon Raphael 1504-06 oil on wood;
St. George killing the Dragon by Peter Paul Rubens, 1620.

Stories like these abound. Yes, some of them have probably been augmented. Some may actually be mostly myth. But enough commonality exists among them that they are undoubtedly based in something that was seen, something that resembled dinosaurs.

According to Genesis all land-dwelling animals were created on Day 6 of Creation Week, along with man. That means that man did dwell with dinosaurs in recent history! These animals would have included beasts of all kinds, reptiles, amphibians and mammals along with those things that slithered, wiggled, ran and crawled. Remember that the word dinosaur was not coined until the early 1800s. These beasts were called dragons before then. Modern paleontology has done a great disservice to these accounts by pounding into the minds of man that dinosaurs evolved over hundreds of millions of years, gave rise to mammals and then went extinct 65 million years ago. Today it is considered both silly and unenlightened to believe otherwise. The evolutionary idea that dinosaurs died out long before man appeared seems to be due to 3 things:

a) The indoctrination that began in the 1800s stressing the millions of years of earth history

b) The Enlightenment view of religion as myth and consequently the view of authentic historical accounts of dragons, especially those contained in the Bible, were also myth

c) The rarity of historical modern eye witness accounts

Nevertheless, the Scriptures are clear in their records that all land creatures were created on Day 6 of Creation Week and then most of them were destroyed 1,600 years later as a result of the Genesis Flood. The only land-dwelling creatures to survive were those who were on the ark. This explanation seems highly plausible for the sudden disappearance of the dinosaurs. The idea that an asteroid caused the selective death of the dinosaurs is just too hard to believe.

Our modern classification system

During the 1700s a man by the name of Carl Linnaeus (1707-1778), a Swedish botanist (one who studies plants), physician, and zoologist (one who studies living animals), came up with a way to organize living plants and animals. His system has been called the Linnaeus Classification System. This system uses one of the science languages, Latin, to help organize plants and animals. Linnaeus' system was used to organize *living* plants and animals. The application of the Linnaeus system to dinosaurs and other extinct creatures came later. After all, it is hard to classify dead things. You cannot observe their eating or living habits and you cannot observe their breeding habits. Carl Linnaeus was a Christian and believed in the Bible. He believed that God had created everything and therefore we could study it and, like Adam, organize God's creation. Linnaeus' system came about a hundred years before Charles Darwin's published ideas and certainly did not have evolution in mind when he devised his system.

During this time a rift began to appear in our culture. Man began to stop believing in God and His Bible and started to trust his own ideas. Linnaeus' classification system began to be used to organize the fossils and dinosaurs. But because fossils are only the stone remnants of once-living things, classification of these things is extremely difficult. This would open the door to a lot of subjective opinion. And because they are the opinions of men, they have changed time and time again through the years.

So, it is time to go back to Genesis. In Genesis 1 God uses the word, *kind* to describe living things. And then God describes what He means. The kinds were to "reproduce after their kind." So, a good definition of kind might be that plants and animals be able to have *offspring*. But since no one has described just how dinosaurs did that, we must guess. Scientists think that they know, but they are just guessing. They don't tell you they are guessing, and this confuses things.

I think there are some things that might allow us to guess as to the kinds of dinosaurs that God had created. Of course, we don't know for sure but we can guess. And I am telling you that I am guessing so you won't be fooled.

When I was a kid, my favorite dinosaur was Triceratops (pronounced *tri-SAIR-uh-tops*). Do you have a favorite dinosaur? Triceratops had a shield of bone that covered his head, and three horns. His skeleton looks like this:

Triceratops

Triceratops' name in the Greek language means *three-horned face*. If you take away the *tri*, then you have *ceratops*, meaning *horned face*. There are other dinosaurs that share part of Triceratops' name. For example, Protoceratops (pronounced *pro-toe-SAIR-uh-tops*. His skeleton looks like this.

Protoceratops

Notice that Protoceratops has a bony shield covering his head, but no horns. Yet, he still has the word *ceratops* in his name. But since we cannot know much about Protoceratops, we have to guess. But let's guess based on God's Bible. That would be the safest thing to do because we do not know for sure. Scientists think they know for sure, but they are just guessing. They do not tell you they are just guessing. What do scientists think? The word Protoceratops means, *first horned face*, or *before the horned face*. Scientists think that the first *ceratops* had no horns at first and then he changed over time to evolve three horns. But again, they are just guessing. And their guess is based on an idea that is not found in God's Bible. God did not create things to evolve or change into completely different things. He created them to reproduce after their kind.

So, what could we say about this dinosaur that is based on God's Bible? Again, we are guessing, but Protoceratops could have been a young Triceratops that had not grown horns yet. Or perhaps he was a completely different kind of dinosaur. We will probably never know.

Scientists call these kinds of dinosaurs, *ceratopsians* (pronounced, *sair-uh-TOP-see-en*, meaning *horned face*). This name would put all the *horned face* dinosaurs into one kind. Here is a picture of the variety of ceratopsians that have been discovered.

Skulls of ceratopsian dinosaurs – Were these evolving creatures or simply variations within the ceratopsian kind?

Since we cannot work with living ceratopsian dinosaurs, we must guess. We can safely guess that these dinosaurs were simply different varieties of the ceratopsian kind. Some of these dinosaurs looked pretty strange.

Now, how would Noah choose a mother and father for the ark? Well, if the ceratopsian dinosaurs were all the same kind, then all Noah would have to do is make sure he chose a mother and a father. God would take care of the rest.

That is exactly what God did with Noah's family. There were eight people on board the ark. But another way to look at that is that there were four couples or four mothers and four fathers. They would have had children after the Flood was done. All Noah would have to do to make sure the ceratopsian dinosaurs could have offspring would be to select a mother and father ceratopsian from the horned face dinosaurs. The variations could have come from the common gene pool that ceratopsians shared.

How many kinds of dinosaurs were there? Well, we don't know for sure, but again, we can guess. Let's use the classification names that paleontologists have used. In that case we have the following kinds of dinosaurs:

1. The **therapods** (pronounced *THAIR-uh-pod)*, whose name means *beast foot*. So all the dinosaurs had feet like this:

Do you recognize them now? Yes, they are like Tyrannosaurs Rex.

2. The **ceratopsians** (pronounced, *ser-a-TOP-see-un)*, which means *horn-faced*. Triceratops is a ceratopsian.

3. The **sauropods** (pronounced, *SORE-uh-pod)*, whose name means *lizard foot*. But that does not make sense, because the sauropods did not have lizard feet! That is silly! The sauropod feet looked like the first picture and lizard feet look like the next picture:

For the sauropods I like the name *long necks* better. And, there were lots of different long-necked dinosaurs.

4. The **hadrosaurs**, (pronounced *HAD-druh-sore*), whose name means *thick, bulky lizard.* That does not help very much does it? I like the name *duck bill* dinosaur. And there were a lot of different duck bill dinosaurs. Now, they don't really have duck bills, but their faces kind of look like duck faces.

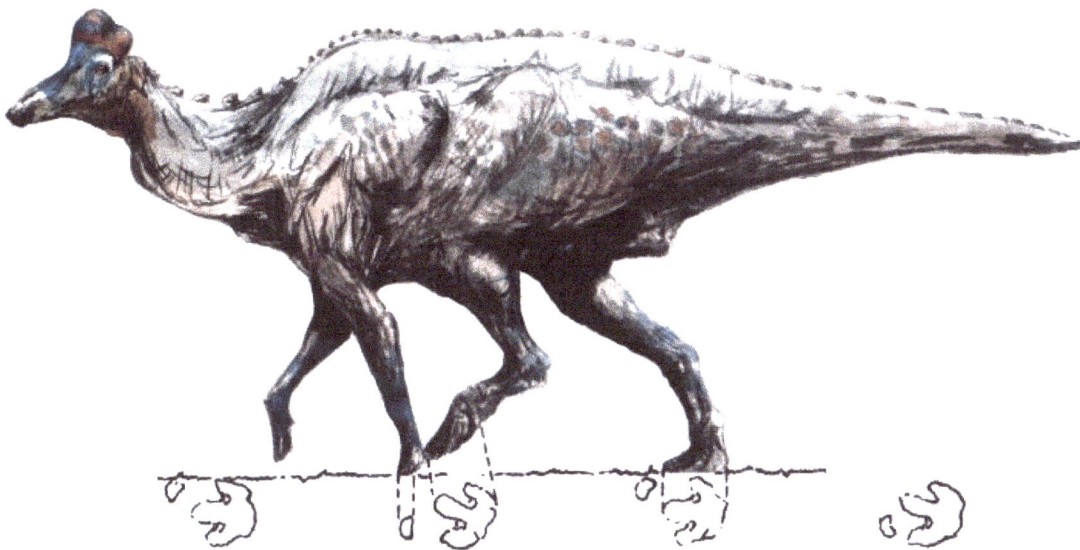

5. The **stegosaurs**, (pronounced *STEG-uh-sores*), whose name means, *roof lizard*. But that does not help much, does it? I like the name, *plated dinosaur*. These were the dinosaurs that had large bony plates along their back.

6. The **pachycephalosaurs**, (pronounced, *pak-ee-SEPH-uh-lo-sores*), whose name means *bone-headed lizard* or *the bone heads*. Funny name for a dinosaur, isn't it? But it describes him well.

7. The **thyreophora**, *(pronounced, thigh-er-OFF-or-uh)*, whose name means *shield bearer or, armor bearer.* This describes the dinosaurs like ankylosaurus, *(pronounced, an-KY-lo-SORE-us)*. These were dinosaurs that looked a lot like the armadillo of today.

So these are seven kinds of dinosaurs that Noah could have worked with. Most of these dinosaurs were seen by human beings throughout the years since the flood. This means that they must have been on Noah's ark in order to have escaped the Flood. The ark and the flood of Genesis 7-8 can adequately explain the tough times that would have followed a global catastrophe such as this. Many creatures apparently did not survive.

Modern paleontology and its funny classification system of dinosaurs
Ever since The Enlightenment view that Genesis was no more than myth, like the writings of other ancient cultures, modern paleontology has aggressively sought a way to restructure the evolutionary history of dinosaurs. This blindness has produced lots of conflicting stories, many of them very funny.

The secular classification of dinosaurs:

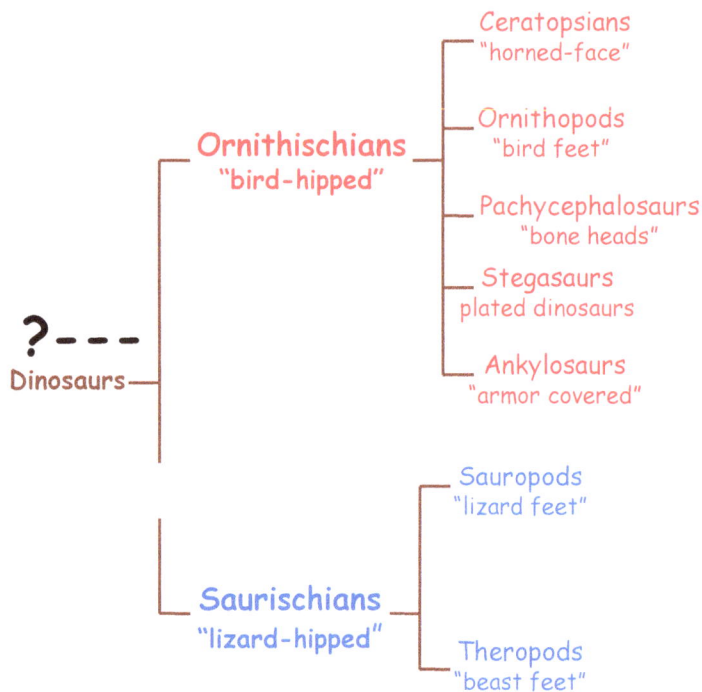

Above is a simple classification chart of dinosaurs according to paleontologists. One of the obviously missing parts is the origin of dinosaurs – from where did they come? Paleontologists have no idea of where these beasts came from. Many thoughts have stirred endless debate as to what ancestor could have possibly given rise to these creatures that have captured the minds and hearts of many thousands of children worldwide.

Let's look at the above modern classification system a little closer and try to get a picture of just how modern paleontologists think about dinosaurs. Let it be stressed at this point that the fossils that have been discovered over the past 250 years have shown only that these beasts:

1. Were different
2. Were rather large, although many were small
3. Possessed lots of variation
4. Have either gone extinct or are in the process of going extinct

Paleontologists divide dinosaurs into two groups – The **ornithischians** (pronounced, *or-nith-ISH-ee-uhns*) and the **saurischians** (pronounced, *sore-ISH-ee-uhns*). All dinosaurs are organized under one of these two groups of dinosaurs. There are two funny parts of this arrangement. One of them lies in the meaning of these names, having to do with *hips* and *feet*. The other one lies in the idea as to which one of these two groups birds supposedly evolved from. Let's continue.

Ornithischian dinosaurs

The word ornithischian means *bird-hipped* and includes all dinosaurs thought to include this particular kind of hip structure. This group includes the horn-faced, the bird-feet, the bone-headed, the plated and the armored dinosaurs.

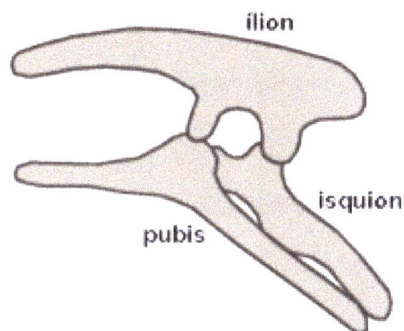

Pictures depicting the ornithischian hip and the dinosaurs thought to possess it.

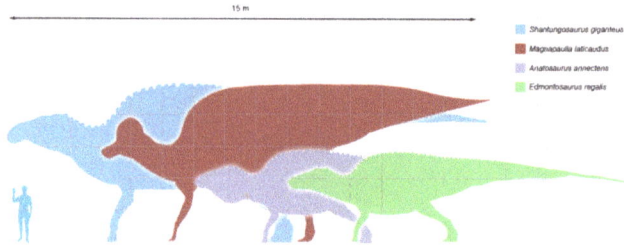

Ornithopods

The ornithischian dinosaurs from top to bottom – the ceratopsians (*horn-faced*), the ornithopoda (*bird-footed*), the pachycephalosaurs (*bone-heads*), the stegosaurs (*plated dinosaurs*) and the ankylosaurs or the thyreophora (*shield bearers*, often known simply as *armored* dinosaurs).

Now here is one the confusing but funny parts to this arrangement. Ornithischians do not have bird hips! They had their own hip structure. And birds have their own hip structures.

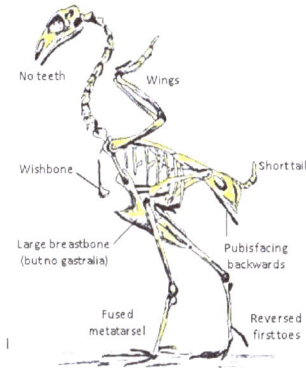

A bird's thigh does not move substantially from its nearly horizontal position, where it provides rigid lateral support to the thin-walled air sacs of the respiratory system. Birds are *knee walkers*, not *hip walkers*. This poses a seemingly **insurmountable** barrier to the dino-to-bird theory. Morris, John D., Frank J. Sherwin. *The Fossil Record.* Institute for Creation Research, Dallas. 2010.

Furthermore, those dinosaurs within the ornithischians called ornithopods (*bird feet*), do not have bird feet! They have their own foot structure.

The comparison of modern bird feet and ornithopod feet – Whatever similarity there might be is not due to evolutionary ancestry, but to a common design and Designer.

Now, here is the other funny part of dinosaur classification. According to evolutionary ideas, dinosaurs were supposed to have evolved into birds. In fact, in today's paleontology circles, birds are the only surviving remnant of the dinosaurs! Many paleontologists in fact call them living dinosaurs. Be that as it may, with all the discussion of birds, which one would you think evolved into birds? Well, I would guess that paleontologists think they would have come from the ornithischians.

After all, the word ornithischian means *bird-hipped*, right? Wrong! Paleontologists believe that birds evolved from the next group of dinosaurs, the saurischians! What?!

Let's take a look at the saurischian dinosaurs.

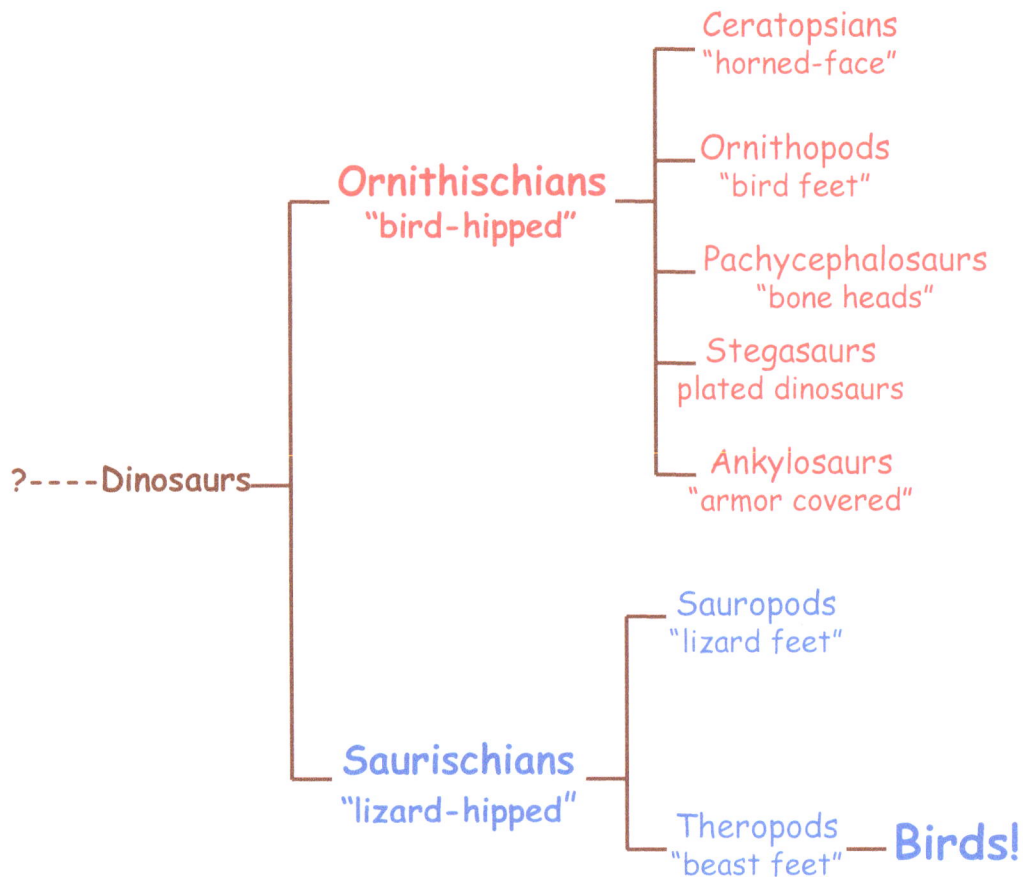

```
                                                    Ceratopsians
                                                    "horned-face"

                                                    Ornithopods
                                                      "bird feet"
                        Ornithischians
                         "bird-hipped"              Pachycephalosaurs
                                                      "bone heads"

                                                    Stegasaurs
                                                    plated dinosaurs

?----Dinosaurs                                      Ankylosaurs
                                                    "armor covered"


                                                    Sauropods
                                                    "lizard feet"

                        Saurischians
                         "lizard-hipped"
                                                    Theropods   — Birds!
                                                    "beast feet"
```

Nurre, Patrick. *Secular Classification of Dinosaurs.* 18/2/2014.

Saurischian dinosaurs

The word saurischian means *lizard-hipped* and includes all dinosaurs thought to include this particular kind of hip structure. This group includes the sauropods, (*lizard feet,* often referred to as the long-necks) and the theropods, *beast feet.* With such an emphasis on the feet of these beasts, you would think these dinosaurs had more to do with a similarity in feet, not hips. Now, dinosaurs do not have lizard hips, for they had their own hip structure. Similarly, lizards have their own hip structure.

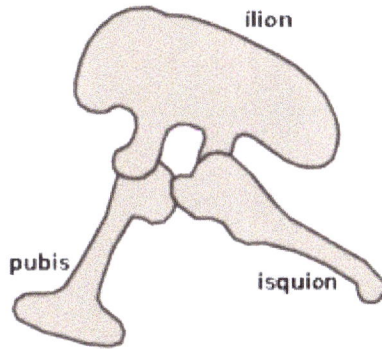

Pictures depicting the saurischian hip and the dinosaurs thought to possess it.
The theropod dinosaurs on the left and the sauropod dinosaurs on the right are both members of the saurischian dinosaurs.

A lizard's hip structure comes out and away from its body, unlike the theropods and sauropods.

Now if you thought you were confused up to this point, wait until you read this! Paleontologists believe that birds evolved from the saurischian dinosaurs, specifically from the theropods, not the ornithischian dinosaurs that possess the word *bird* in their name! Are you totally confused yet?

Would it not be more scientific to simply acknowledge the kinds of dinosaurs and the variations within the kinds? Would it not be more scientific to simply acknowledge

103

the bird kind as a separate and unrelated kind from the dinosaurs? Would it not be more scientific to simply note the similarity in skeletal design as a mark of the Designer?

Even more confusing is a report that came from the April 2000 Florida Symposium on Dinosaur-Bird Evolution, reported by Jonathan Wells who was in attendance at this Symposium. On the second day of the symposium, William Garstka reported that he and a team of molecular biologists from the University of Alabama had extracted DNA from the fossil bones of a 65-million-year-old dinosaur. (Evidence from other studies would suggest that DNA older than a million years cannot yield any useful sequence information.) His team compared this DNA to known DNA from other animals and found it to be most similar to birds. What an exciting find! They concluded that they had found "the first direct genetic evidence to indicate that birds represent the closest living relatives of the dinosaurs!" Their report was reported the following week in *Science* magazine.[14]

But there is a confusing part to this report. The DNA Garstka and his colleagues recovered was 100% identical to the DNA of a living turkey – a bird! Was this the proof that evolutionists were looking for? It sure seems like it was. So, what dinosaur did Garstka extract DNA from? The DNA was supposedly recovered from a Triceratops! Triceratops is grouped as an ornithischian dinosaur! What is so fantastic about this? According to paleontologists, birds were supposed to have evolved from *saurischians* not from the ornithischians to which Triceratops belonged! What is even more amazing from the report on the Symposium, is that no one seemed to be rattled by this. So long as it could be established that birds came from dinosaurs, the evidence, which was contrary to standard paleontological thinking, did not seem to matter. This is very disturbing, because if this is true, then the whole idea that birds evolved from theropods would go down the drain! Some who were there wondered if the whole thing was just a hoax. Even Garstka wondered if he was the victim of a prank. Time will tell. One thing is certain – most of the scientists who attended the symposium believed the whole affair to be entirely credible!

Were Dinosaurs on the Ark of Noah?

Many people make fun of the story of Noah's ark today. They don't think it is a true story. But did you know that Jesus believed in the story? Read the story from the gospel of Matthew 24:35-40. Was Jesus a nutcase? Did our Savior believe in stupid myths like Noah and his boat? Did this Jesus, the Son of God who created the dinosaurs, believe in fairy tales?

---

[14] Holden, Constance. 1 Apr. 2004. Web. 9 Jan. 2015. ‹http://news.sciencemag.org/2000/04/dinos-and-turkeys-close-relatives›.

When most people think of Noah's ark, they think it looked like this picture or something like it:

But in the Book of Genesis 6 we are told that Noah was to build an ark (the word ark means, *box*) that was to be 450 feet long, 75 feet wide and 45 feet high with three floors! That is a huge box. Just how big was this ark?

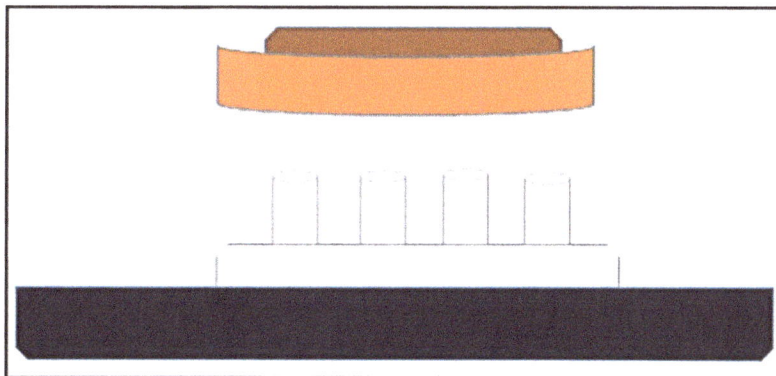

Noah's Ark compared to the Titanic!

It is clear from the Book of Genesis that the ark was meant to carry lots and lots of animals and food. We have already seen that the Flood lasted well over a year. Noah would have needed this huge ship to carry, house and feed all the animals.

What kind of animals did Noah take onto the ark? Genesis chapter 6 and 7 tell us that Noah took two pairs of every land-dwelling, air-breathing kind of animal. With some of the animals he took seven pairs, those that God told him were clean. I am not sure what those were, but God wanted Noah to take enough of every *kind* of animal to keep them alive to produce offspring after the Flood.

What about dinosaurs? Some of the dinosaurs were huge! Well, if I was Noah, I would not take the old dinosaurs, those that were huge and would eat a lot of food. I would take the young dinosaurs. Dinosaurs were very small when they were born. After they had lived for a while, they grew to be very large.   The picture below shows how big a Hadrosaur egg was – about the size of an ostrich egg. The next picture shows a 35-foot long Hadrosaur. Obviously, he must have been smaller when he was born!

A Hadrosaur egg on the left is compared in size to a pencil. The picture on the right represents an adult Hadrosaur, about 35 feet long. Obviously, it was much smaller when hatched.

Dinosaurs seem to have grown very much the same way that crocodiles grow today! Unless crocodiles die, they will keep growing and growing. But they are small when they are born.

Noah might have taken 7-12 different pairs of parent dinosaurs on board the ark. The ark was large enough to hold all the kinds of animals that lived in Noah's day. The ark was also large enough to hold food and water and Noah's family.

How do we know that Noah took dinosaurs on board the ark?
We know this, because people, who lived after Noah, reported seeing them! In order to have escaped the flood, these animals would have needed to be on the ark. One of those people was Job. Job lived a short time after the Flood. In the Bible, Job was told by God to think about two beasts He had made. He did this to impress Job with how powerful God was. If God had created creatures that were large enough to scare man, then God must be pretty powerful. Job chapters 40-45 describe two beasts that were very, very big!

Behemoth
The first beast that God described to Job was a creature called, Behemoth, (*pronounced, buh- HEE-muth*). The way the creature is described makes him look like a large sauropod dinosaur.

In Job 40:15-24 God describes a beast to Job that fits one of the sauropods like diplodocus above. "He bends his tail like a cedar; the sinews of his thighs are knit together. His bones are tubes of bronze; his limbs are like bars of iron."

The second beast that God described to Job was a creature called, Leviathan, (*pronounced, le-VIE-uh-thun*). The way this creature is described makes him look like either a Mosasaur or a Megalodon – huge creatures of the sea.

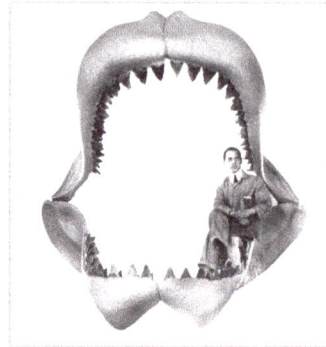

Mosasaur skeleton and Megalodon jaw.

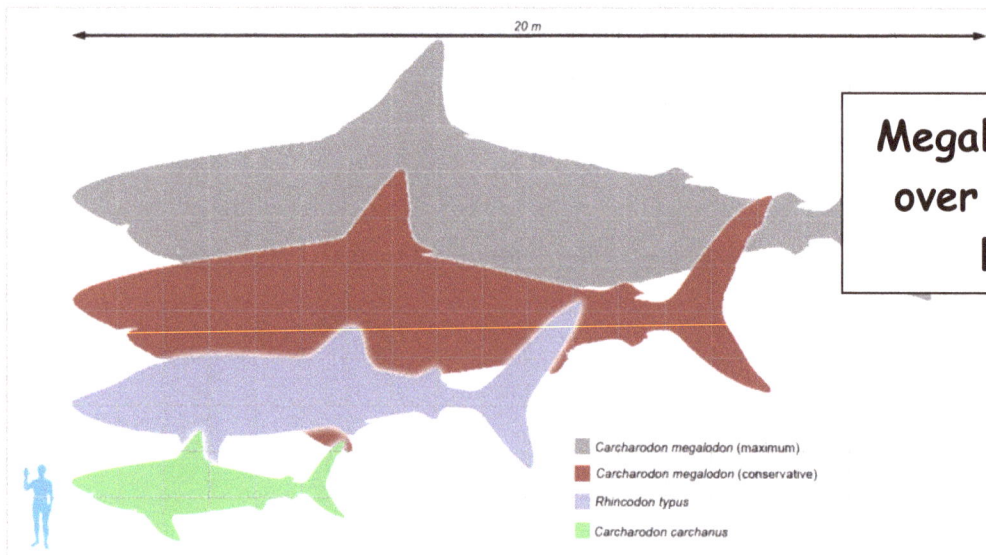

Megalodon was over 54 feet long!

Carcharodon megalodon (maximum)
Carcharodon megalodon (conservative)
Rhincodon typus
Carcharodon carcharius

Megalodon Tooth

Job 41:1-34, "...Will you be laid low even at the sight of him? ...No one is so fierce that he dares to arouse him; who then is he that can stand before Me? ... Around his teeth there is terror." You may have noticed different colors of the teeth. This is due to different chemicals in the environment in which they were buried.

Many people are familiar with the story of St. George. A story was widely circulated during the Middle Ages that he bravely killed a *dragon*. Was it a dinosaur that he killed? People knew enough about dragons to report having seen them. I wonder if these could have been the *dragons* descended from those which had been on Noah's ark?

St. George, the dragon slayer: a popular story of the Middle Ages

Why don't we see more dinosaurs today?

After the Flood was over, in Genesis chapter 9, God told Noah that not only could he eat plants, but now he could eat animals too. With that in mind, I wonder if man began to hunt dinosaurs for sport and for food. This could have had a harsh effect on the survival of certain animals. Dinosaurs might have been hunted to extinction!

A second reason might be the dramatic change in weather that came on the Earth after the flood. With the large number of volcanoes that were erupting during and after the flood, ash clouds would have kept the warmth of the sun from heating the earth effectively. More moisture from the new oceans might have risen into a cooler atmosphere, causing a lot of snow to fall. This would have brought on an *Ice Age*. We know that millions of mammoths were caught in this weather change all across Siberia and parts of Alaska. Dinosaurs might not have done so well with this extremely uncomfortable weather. Life was also tough for man after the flood. After the Tower of Babel, about 100 years after the Flood, when God scattered man, many must have looked for caves as their homes. Many who had relied on other people to help them build things, no longer could. Some tribes of people even died out during this time. People like the Cro-Magnon people and the Neanderthal people might have not survived the harsh weather in the north. (We will talk more about these people in a later lesson). Dinosaurs might not have done so well either.

Before the Flood, the fossils of dinosaurs and plants seem to indicate that the dinosaurs were used to living in warm tropical climates. After the Flood, that all changed for hundreds of years until the weather began to warm up again. By then it might have been too late. Dinosaurs and cold weather just don't seem to go together!

Although most pictures of the woolly mammoth show him living in cold weather, this is not known for sure. After all, he is grouped with the elephant kind! And elephants live in the jungle, not in Siberia. In fact, food found in frozen mammoths' stomach show that it was tropical. The woolly mammoth might simply have been a variety of elephant that had long hair. What we do know for sure is that millions of them seemed to have frozen to death in the Siberian region as it was changing rapidly after the flood.

Woolly Mammoth

Dinosaur-to-Bird Evolution
It was Thomas Huxley, who in 1861 proposed that Archaeopteryx evolved from a theropod dinosaur (compsagnathus). Dinosaur-to-bird evolution has remained firmly entrenched in paleontology as well as in the mind of the public ever since then.

Thomas Huxley (1825-1895), Compsagnathus skeleton, Archaeopteryx

110

When I was a kid, there were two main imaginative ideas of how dinosaurs could have evolved into birds. One idea proposed that a dinosaur-like ancestor climbed a tree and with repeated attempts at jumping out of the tree, developed first the ability to glide and then wings to fly. This idea has become known as the *top-down* idea. Another idea proposed was that as the need for food increased due to extinction of certain insects or some other factor, a dinosaur began flapping its claws wildly in order to catch flying insects. Through a series of the right kinds of evolutionary changes the dinosaur eventually took off and flew. This idea is known as the *bottom-up* idea. Now, obviously paleontologists believe the stories are a little more complicated than this. But these ideas are still being taught today.

It normally escapes some biologists' attention at this point, but these ideas are leftovers from a thoroughly discredited teaching of evolutionists of the 1800s – *acquired characteristics*! This idea stated that a creature, the giraffe for example, by stretching its neck for food in the upper branches of trees, could pass on this characteristic to its offspring. And with repeated generations of doing this, developed the long necks we see today. Darwin sincerely believed this idea. Today we know that changes in genetics would be the only way to produce totally different kinds of anatomical structures.

The two views of dino-to-bird evolution – science or imagination gone awry?

One of the clearest examples of scientific speculation today is in the ceaseless and tireless effort that has been expended on proving that birds evolved from dinosaurs. Obviously if scientists are right in teaching that birds evolved from dinosaurs, then the Bible clearly cannot be true! So, we must examine this idea.

Cladistics

A new idea for classifying animals surfaced in the 1950s - *Cladistics* - classification via outward or perceived shared characteristics. This has become the system that most biologists use today. The biologists who disagree, point out the subjectivity of the whole system. In fact, as we shall shortly see, even the fossil evidence has been given second place to cladistics. Let me explain.

In Darwin's day and for the next century the firmly held belief was that all living things descended from some common ancestor through slow and gradual transitions over a period of hundreds of millions of years. And it was believed that after enough fossil collecting, missing links would be found. Cladistics puts aside this belief and states that the fossils are secondary to the *idea* of evolution. Since the fossil record is so incomplete, the fossils must be interpreted to give the answers evolutionists look for. In other words, the predominant belief is that birds evolved from the theropod dinosaurs. Therefore, the evidence that is sought is that which demonstrates this. A system of matching the most outward characteristics that are shared among candidates is arranged to demonstrate the perceived scientific fact of evolution. Therefore, the fossils should show fur or feathers. Even if it means manufacturing them, feathers are part and parcel of a theropod's makeup because birds share the most outward characteristics with theropods. The fossils are simply window dressing. A good example of this is much heralded bambiraptor.

Bambiraptor

Bambiraptor - the skeleton, the fossil, and the reconstructed dinosaur. Although no feathers were found on the original fossil, cladistics demanded it - it ought to have feathers, because after all, it is (believed to be) the ancestor of birds.

Another example of cladistics applied concerns the case of **Deinonychus.** The word Deinonychus (*die-NON-i-kus* or, *die-no-NY-kus*), according to Jack Horner of the Museum of the Rockies, means, *terrible claw*, after the sickle-shaped claw on its feet. Discovered in the 1960s by John Ostrom, the dinosaur was originally drawn to look like the first picture below, a reasonable interpretation of the bones. But then the philosophy of cladistics got hold of him and immediately Deinonychus went through a transformation from reptile-like dinosaur to bird-like dinosaur. Because of the belief that theropods gave rise to birds, theropods must have existed with feathers.

An artist's rendition of the original Deinonychus based on the fossil found in the 1960's, on the left; the modern artist's rendition on the right, from the Museum of the Rockies in Bozeman, Montana. Deinonychus now is pictured with feathers and fur to enhance the now-believed, thoroughly entrenched idea that dinosaurs gave rise to birds. It does not matter that none of the official candidates were found as fossils with feathers, the idea is what is important. The science has been made to fit the idea.

One of seven fossils of Archaeopteryx that have been found. The fossil very clearly shows feather and wing imprints. Archaeopteryx is considered to be fully bird today, extinct, yet fully bird. He is no longer considered to be the evolutionary missing link as it was first described in the 1800s.

The Strange Case of The Evolution of the Wing and The Path to Birds
A good illustration of how cladistics has taken over evolution and reason is found in this beautiful depiction called, *The Path to Birds*, from the National Geographic Society.

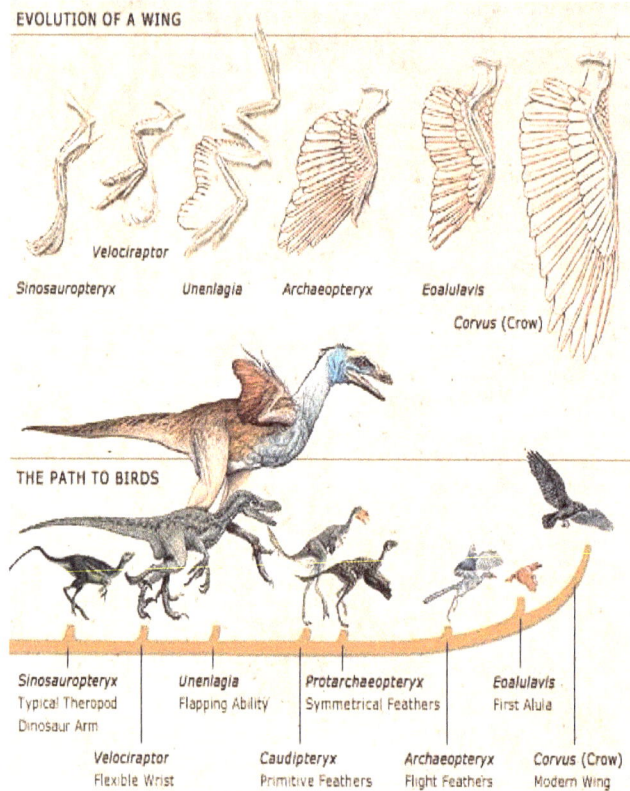

Although an elaborate series of pictures convincingly presents the evolution of the wing and birds, there is no clear evidence available. It is all conjecture. Notice that in this supposed evolution of the wing, there is no evidence and no explanation of how or why scales would turn into feathers. It is treated as fact that they just did. Also, as you will see in the picture below, the candidates are all out of order chronologically!

The picture very attractively presents the history of how birds, wings and flight evolved from dinosaurs. If you have ever heard of the saying, "A picture is worth a thousand words," this presentation certainly accomplishes that. The candidates are nicely drawn and placed in such a way as to create a scientific history of flight and the evolution of birds with drawings of wings and how they developed over millions of years of evolutionary change. Just how accurate is this presentation? A cladist (one who practices cladistics) would not really care if it accurately represents the fossils or not. The idea is deemed true and is the most important thing. So the evidence must be made to fit the idea. I want to make it clear that the paleontologists who constructed this display are aware of everything I am going to show you, but they are blinded by their belief in evolution and especially in the belief that birds evolved from dinosaurs.

In order of appearance in picture, *The Path to Birds*, are the following candidates for dino-to-bird evolution:

1st - **Sinosauropteryx** – meaning *Chinese reptilian wing*
2nd - **Velociraptor** – meaning *swift seizer*
3rd - **Uenlagia** – meaning *half bird*
4th - **Caudipteryx** – meaning *tail feather*; now classed as a flightless bird
5th - **Protarchaeopteryx** – meaning *before Archaeopteryx*
6th - **Archaeopteryx** – meaning *ancient feather or wing*
7th - **Eoalulavis** – meaning *dawn bastard-wing bird*; now classed as an extinct group of primitive birds
8th - **Corvus** – Latin for *raven or crow* (modern wing)

So, what's the problem? Take a look at the picture again, now with the geological dates of the rock layers in which the critters were found. These are the uniformitarian ages given for these candidates.

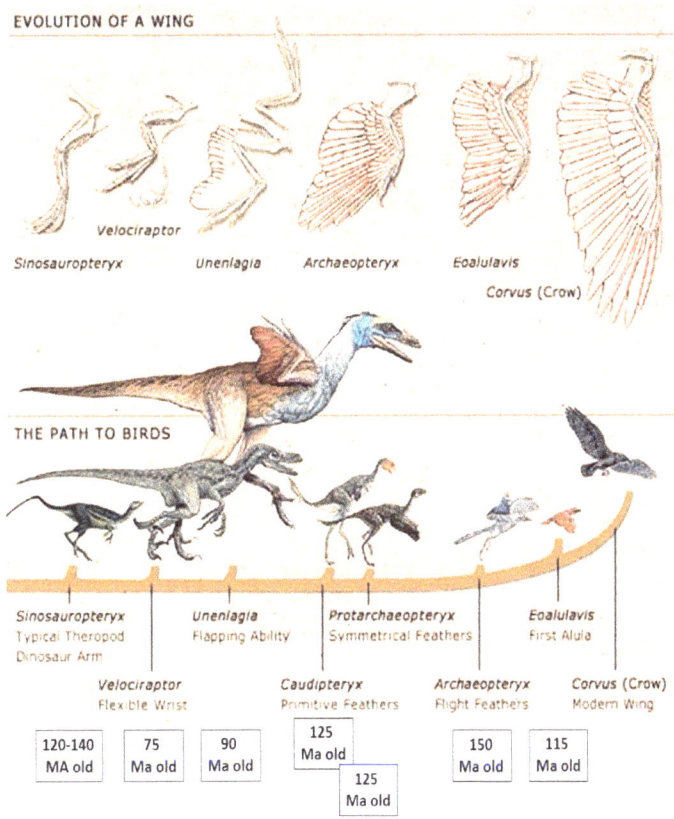

EVOLUTION OF A WING

Velociraptor

Sinosauropteryx   Unenlagia   Archaeopteryx   Eoalulavis

Corvus (Crow)

THE PATH TO BIRDS

Sinosauropteryx		Unenlagia		Protarchaeopteryx		Eoalulavis	
Typical Theropod		Flapping Ability		Symmetrical Feathers		First Alula	
Dinosaur Arm							
	Velociraptor		Caudipteryx		Archaeopteryx		Corvus (Crow)
	Flexible Wrist		Primitive Feathers		Flight Feathers		Modern Wing

| 120-140 MA old | 75 Ma old | 90 Ma old | 125 Ma old | | 150 Ma old | 115 Ma old |
| | | | | 125 Ma old | | |

As you can see, the chronology is all out of order. Why aren't the dates printed on the original poster? It would seem that dates and fossils are really not the important thing, the idea is!

The *oldest* candidate pictured above is Archaeopteryx, an extinct but fully flying bird. The rest of the candidates should be older than Archaeopteryx in order to qualify as ancestors. The display is befuddling from a scientific point of view. The actual evidence points to backward evolution, not to *The Path to Birds!* Why would a highly respected magazine purposely display an idea that has no scientific basis and wrong scientific evidence? You can see how the "science" of cladistics dominates scientific thinking today.

Using our Biblical Geologic Column, we can see where most of the dinosaurs would have undergone extinction. Those that were on the ark died out as time went on. There are two reasons for this: (1) the subsequent harsh climate after the Flood, and, (2) possibly from being hunted to extinction by man.

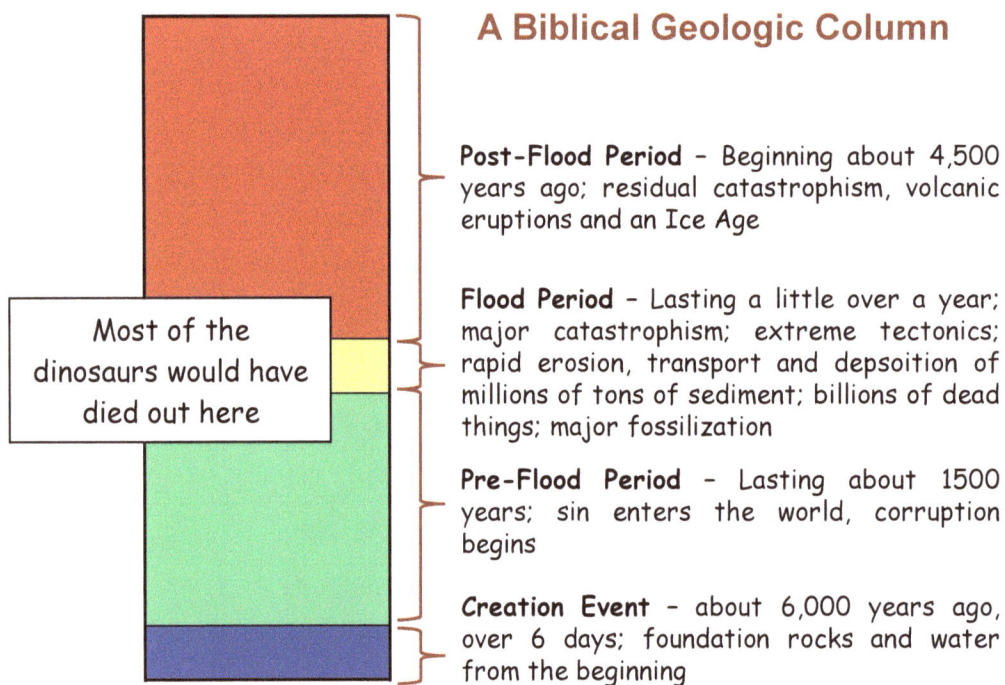

A Biblical Geologic Column

Most of the dinosaurs would have died out here

Post-Flood Period – Beginning about 4,500 years ago; residual catastrophism, volcanic eruptions and an Ice Age

Flood Period – Lasting a little over a year; major catastrophism; extreme tectonics; rapid erosion, transport and depsoition of millions of tons of sediment; billions of dead things; major fossilization

Pre-Flood Period – Lasting about 1500 years; sin enters the world, corruption begins

Creation Event – about 6,000 years ago, over 6 days; foundation rocks and water from the beginning

Go to a library and obtain a copy of the November, 1999 issue of *National Geographic*. Read the story of Archaeoraptor. What kinds of things do you notice about the bias behind the story?

Please take Quiz #7, Appendix B

Lesson 8 – Transitional Fossils

The Keys to Evolution

Words to know: *transitional fossil   variation*

What is a **transitional fossil**?   Darwin's ideas of evolution were that species had changed gradually over hundreds of millions of years through many intermediate steps.   He had tried to demonstrate this by observing and recording the subtle differences in the beaks of the finch.

1. Geospiza magnirostris.
2. Geospiza fortis.
3. Geospiza parvula.
4. Certhidea olivasea.

Darwin reasoned that subtle changes in beak size and shape could, with millions of years and environmental pressures, change into something entirely new.   There had been a popular belief among many scientists of the time that the Bible taught *fixity of species*, that species could not change; God had fixed them, and so evolution would be prevented from happening.   This was an *interpretation* of the word *kind* in Genesis chapter 1.   It was an unfortunate interpretation because everyone could point to **variation** within dogs and other animals.   Darwin proved that species do change.   But what Darwin actually demonstrated was that they only change *within* species.   Darwin took it one step further.   Because species change, then over millions of years these

118

changes could multiply and change into new kinds of living things. Therefore, the Bible must be wrong. He failed to see that it was the interpretation that was wrong, not the Bible. Evidence from nature shows us that there is limited variation within kinds of living things, but not unlimited variation so that totally new kinds are produced. For evidence of his ideas Darwin looked to the fossils, as they were the record of past life.

Fossil collecting was a young science in those days. Paleontology did not exist as a scientific discipline. There were no vast collections of fossils to study. Therefore, the fossil evidence would need to simply be sought and organized to ultimately prove Darwin right. But even in his day, Darwin was bothered by the lack of transitional fossils. Transitional fossils are those that show some sort of evolutionary transition between one kind of species and another totally different kind of species. Darwin commented on this obstacle in his book, *On the Origin of Species*.

> Why is not every geological formation and every stratum full of such intermediate links? Geology assuredly does not reveal any such finely-graduated organic chain; and this is the most obvious and serious objection which can be urged against the theory.[15]

His hope was that as the study of fossils grew, his ideas would be vindicated. At least as late as the late 20th Century, these intermediate fossils had not surfaced. The famous 20th Century Paleontologist, Stephen J. Gould stated, "The extreme rarity of transitional forms in the fossil record persists as the trade secret of paleontology."[16]

Fossils are the only evidence for life in the past. In order to embrace evolution as a scientific explanation for life, we must be able to see how life evolved in the fossil evidence. Just how many steps should the fossils represent? This is an unknown. One thing is certain. We must be able to bridge any gaps between species and that can only be done through the fossil evidence. What kind of fossil evidence have evolutionists themselves claimed as support for Darwin's ideas?

Archaeopteryx
One of the icons of evolution that has served for many years as the showcase for proof of a transitional fossil has been Archaeopteryx, meaning *ancient wing*. The common consensus among most paleontologists is that it is a bird from the Upper Jurassic, about 144-150 million years ago. They believe it is the link between

---

[15] Darwin, Charles. "Chapter 9, On the Imperfection of the Geological Record." *On the Origin of Species by Natural Selection*. 1859. Print.
[16] Gould, Stephen J. "Evolution's Erratic Pace." *Natural History* 1 May 1977: 14. Print.

reptiles and birds. The first Archaeopteryx was found in 1860 near Solnhofen in Bavaria, Germany. To date, seven skeletons have been found, and one feather.

Many paleontologists have said that Archaeopteryx looks like a reptile with wings and feathers: that is, it has a mouth with teeth, claws, and a long tail like dinosaurs or reptiles. Interesting, isn't it, that *today*, when they are born, South American hoatzin have claws on their wings when they are young, just like Archaeopteryx.

If Archaeopteryx is fully bird, then he cannot be a transitional link between reptiles and birds. If the hoatzin is a living bird that looks like Archaeopteryx and is modern, then how can these features be used to show evolutionary transition? Archaeopteryx is considered today to be an extinct bird with some interesting characteristics, more like a mosaic similar to the duck-bill platypus. Dr. Alan Feduccia, a world authority on birds at the University of North Carolina at Chapel Hill and an evolutionist himself said,

> Paleontologists have tried to turn Archaeopteryx into an earth-bound, feathered dinosaur. But it's not. It is a bird, a perching bird. And no amount of 'paleobabble' is going to change that.[17]

And yet when asked for an example of an intermediate link, most evolutionists will cite Archaeopteryx as one of the best examples of a transitional link! If he was a bird, then he could fly which means that he had all the physical components necessary for flight. Who was his ancestor?

Archaeopteryx appears to have been an extinct bird, as far as we know, but a bird nevertheless. As early as 1861 Thomas Huxley, a biologist, known as *Darwin's Bulldog*, proclaimed Archaeopteryx to be a transitional link between birds and theropod dinosaurs. Archaeopteryx has remained such in the minds of most people since then.

---

[17] Feduccia, A.; cited in: V. Morell, "*Archaeopteryx:* Early Bird Catches a Can of Worms," *Science* **259**(5096):764–65, 5 February 1993.

Here is one of the famous fossils of Archaeopteryx with a representation that shows him to be a bird. In the picture of the fossil, you can definitely see the feathers and tail feathers as well as the skeleton of a bird properly built for carrying feathers and for flying.

What I think is amazing is how much liberty artists have taken to portray ideas. Take a look at the different artist's renditions of Archaeopteryx below. Some show him flying. Some show him running, as if he has not quite evolved into a bird yet. Some show him more reptilian to communicate the idea that he evolved from reptiles.

Artists' renditions knowingly or unknowingly communicate evolutionary ideas. These three pictures are each a very different interpretation of the fossil.

Whales

Another claimed evolutionary transition is between land animals and whales. This is rather a farfetched idea, but it is believed by evolutionists to be the case. So we will look at it. Whales are classed as cetaceans. The word cetacean is from the Greek, *ketos*, for whale. These are creatures defined as aquatic placental mammals having no hind limbs and possessing a blowhole for breathing. The group includes toothed whales (dolphins, porpoises, etc.) and whalebone whales. It is quite a large class of aquatic creature which shows typical variation within its group but also similarities among its group. It was on the fifth day of creation that God created sea creatures in Genesis 1:20-23.

121

What do scientists say about whale evolution? The University of Wisconsin tells the story this way.

> Once upon a time, about 50 million years ago, a deer-like mammal called Indohyus waded into shallow water. He might have been fleeing a predator or eating tasty aquatic plants. Today, he's recognized by evolutionary scientists as an ancient relative of modern whales. The fascinating and complex story of whale evolution will be explored in a public lecture... The campus event features an address by Hans Thewissen, an expert on the evolution of whales who in 2007 announced the discovery of the missing link between whales and their terrestrial ancestors with a fossil skeleton of Indohyus from the Kashmir region of India.[18]

The announcement of this event was accompanied by the following poster in honor of Darwin Day, February 2, 2012. Pretty amazing isn't it? Next to the poster is a representation from a Grade 4 unit study on whale evolution depicting the same idea.

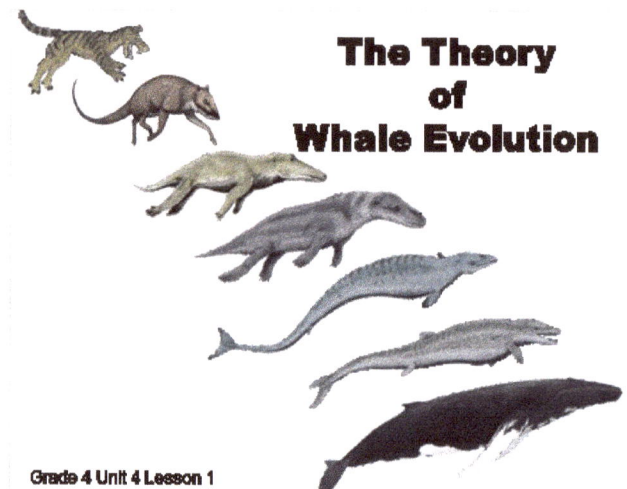

Now let's think about this idea for a minute. What kinds of evolutionary changes would need to take place in order for a land mammal to turn into a whale? Cetaceans have many unique features to enable them to live in water. Consider the following:

- Enormous lung capacity with efficient oxygen exchange for long dives

---

[18] 24 Sept. 2013. Web. 9 Jan. 2015. <http://www.uww.edu/news/archive/2012-02-darwin>.

- A powerful tail with large horizontal flukes enabling very strong swimming
- Eyes designed to see properly in water with its far higher refractive index, and to withstand high pressure
- Ears designed differently from those of land mammals that pick up airborne sound waves, with the eardrum protected from high pressure
- Skin lacking hair and sweat glands but incorporating fibrous, fatty blubber
- Whale fins and tongues have counter-current heat exchangers to minimize heat loss.
- Nostrils on the top of the head (blowholes)
- Specially fitting mouth and nipples so the baby can be breast-fed underwater
- Baleen whales have sheets of baleen. Baleen is a strong, yet flexible material made out of keratin, a protein that is the same material that makes up our hair and fingernails. Baleen hangs from the roof of the whale's mouth and filters plankton for food.

But where are the fossil transitional remains that would show these features developing over time? There are none – absolutely none. When challenged with this, evolutionists simply respond by saying that only certain animals and plants were preserved as fossils out of the many that had existed. This is of course circular reasoning. It is convenient that only those completed or fully evolved animals and plants were preserved.

If you stop and think about it, the kinds of genetic changes needed in the genetic code of the land mammal evolving into a whale would be immense! First, there is no source from which the land mammal would have derived this genetic information or the ability to capture it and synchronize it with the rest of the developing body. Second, there are many significant changes required for a whale to evolve from a land mammal. One of them is to get rid of its pelvis. Land mammals have a pelvis that serves a lot of purposes. But a shrinking pelvis would not be able to support the hind-limbs still needed for walking while the hypothetical candidate was evolving. So the hypothetical transitional form would be unsuited to both land and sea, and hence the evolving animal would be extremely vulnerable, not fitted for either land or sea.

The lack of transitional forms in the fossil record was realized by evolutionary whale experts like the late E.J. Slijper: "We do not possess a single fossil of the transitional forms between the aforementioned land animals [i.e., carnivores and ungulates] and the whales." [19] The lowest whale fossils in the fossil record show they were completely aquatic from the first time they appeared as fossils.

It seems that a special creation is a better explanation of what we observe in the fossil record than the stories we hear in evolutionary philosophy. And yet it is this very philosophy that has a stranglehold on all who wish to keep their faith and graduate from a university with a degree in science.

The Actual Fossil Evidence

Hundreds of thousands of representative creatures and plants from the past have been well preserved in the fossils. The fossil record is rich with fine examples. And these fossils in many cases are exquisitely preserved and record minute details. Certainly, there should be some fossils that record at least some of the transitional changes that would have been a part of evolutionary history!

The major obstacles for evolution would be the genetic information required to change microorganisms to shelled creatures, invertebrates to vertebrates, fish to amphibians, amphibians to reptiles, reptiles to mammals and mammals to man. Of course, there are countless steps of genetic progress that must be accomplished in order for new creatures to survive. It is mind-boggling to think about all the genetic and environmental hurdles that must be overcome in order for new creatures to evolve, thrive and reproduce their offspring. There have been volumes of books written about these necessary changes. The mind of man can *imagine* anything, but the only record of this history that would come close to documenting this progress is the fossil record. We must be able to collect and identify representative fossil specimens that would show the changes demonstrating the long history of evolution. What do the fossils show and what do the experts say?

Microorganisms to shelled creatures

Evolutionists insist that life has been evolving for the last 2-3 billion years. But the first multicelled, shelled creatures appear suddenly in the fossil record according to evolutionists about 550 million years ago. So, where are the fossils that would document the transition of the various living things that preceded these multicelled, shelled creatures? David Axelrod, evolutionist, stated,

> One of the major unsolved problems of geology and evolution is the occurrence of diversified, multi-cellular marine invertebrates in Lower Cambrian rocks on all the continents and their absence in the rocks of greater age...when we turn to examine the Precambrian rocks for the

[19] Sliper, E.J. *Whales and Dolphins.* USA: U of Michigan, 1962. 18. Print.

forerunners of these Early Cambrian fossils, they are nowhere to be found. Many thick (over 5,000 feet) sections of sedimentary rock are now known to lie in unbroken succession below the strata containing the earliest Cambrian fossils. These sediments apparently were suitable for the preservation of fossils because they are often identical with overlying rocks which are fossiliferous, yet no fossils are found in them. Axelrod, Daniel.[20]

What is significant about this statement is that nothing has changed from 1958. In other words, ever since 1958 no explanation has been universally accepted as to why this case persists, and why no fossils have been discovered to show otherwise.

The point is often made that soft-bodied creatures of the Precambrian were not preserved, so that the evolutionary changes would not be recorded from soft-bodied to hard-shelled creatures. But the first picture above shows adequate fossilization of jelly fish in the same time period as trilobites. Non-fossilization is not the issue. Would it not be more scientific to simply conclude with Genesis that "In the beginning, God created…kinds reproducing after their kinds…," and that about 1600 years after the creation the earth was destroyed by a global flood?

Invertebrates to vertebrates

The idea that the vertebrates were derived from the invertebrates is purely an assumption that cannot be documented from the fossil record. F.D. Ommanney, expert in fossil fish says,

How this earliest chordate (fish) stock evolved, what stages of development it went through to eventually give rise to truly fishlike creatures, we do not know. Between the Cambrian when it probably originated, and the Ordovician when the first fossils of animals with really fish-like characteristics appeared, there is a gap of perhaps 100 million years which we will probably never be able to fill. [21]

[20] Axelrod, Daniel. "Early Cambrian Marine Fauna." *Science* 4 July 1958: 7. Print.
[21] Ommanney, F.D. *The Fishes*, Life Nature Library, Time-Life, Inc., New York, 1964, p. 60.

Fish to amphibians to reptiles to mammals

One of the first major transitions that is talked about in evolutionary circles is the development of feet in fish to enable the new creature to leave the water and to walk on land. Certainly, there must be some fossil evidence to support such a notion. However, not a single transitional form has ever been found showing an intermediate stage between the fin of the crossopterygian (fish) and the foot of the ichthyostegid (a so-called amphibian). Although there are close to 500,000 fossil fish specimens in museums all around the world, there are no specimens of transitional forms to document the change from fish to amphibian. Dr. David Raup, paleontologist at the University of Chicago has stated in various interviews that no fish had ever been found with feet and legs.[22]

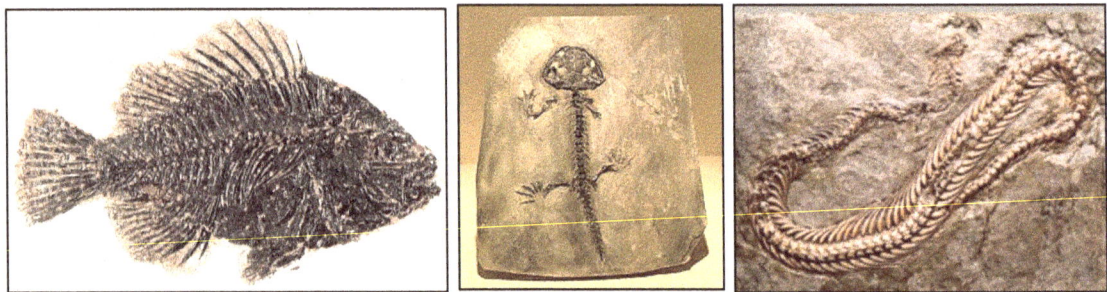

In the fossil record fossil fish are clearly identifiable as fish, fossil amphibians are clearly identified as amphibians and fossil reptiles are clearly identified as reptiles. With such great examples of fossil preservation that abound in the fossil record, certainly some amount of the numerous transitional forms would have been preserved!

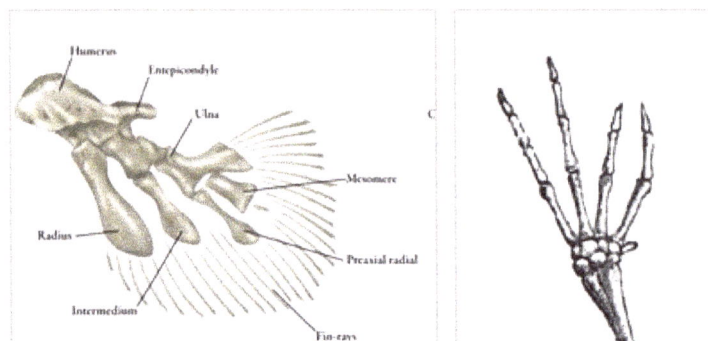

Fish fin   *Amphibian foot*

The transition from fin to feet is no small matter. The fin is laced with rays that give it strength. It is quite different from the bones in the hand or foot. Most importantly, the bones in the fin are not connected to the backbone, a condition that is necessary for walking. The transition appears to be wishful thinking.

---

[22] Interview with Luther Sunderland in <u>Darwin's Enigma</u>, Master Books, 1988, page 73.

Museums house close to 30 million various dinosaur bones. Close to 200,000 complete or nearly complete skeletons have been constructed from these various bones. But the question about what the common ancestor is and the various connecting links within the dinosaurs persists today as a major unsolved mystery. Dr. Weishampel, anatomist and paleontologist at Johns Hopkins University and lead editor of the encyclopedic reference book *The Dinosauria*, stated in a 2005 interview, "From my reading of the fossil record of dinosaurs, no direct ancestors have been discovered for any dinosaur species. Alas, my list of dinosaurian ancestors is an empty one."[23]

As to the transition to mammals, the famous paleontologist G.G. Simpson commented about the fossil forms representing this evolutionary transition,

> This is true of all thirty-two orders of mammals…The earliest and most primitive known members of every order already have the basic ordinal characters, and in no case is an approximately continuous sequence from one order to another known. In most cases the break is so sharp and the gap so large that the origin of the order is speculative and much disputed.[24]

 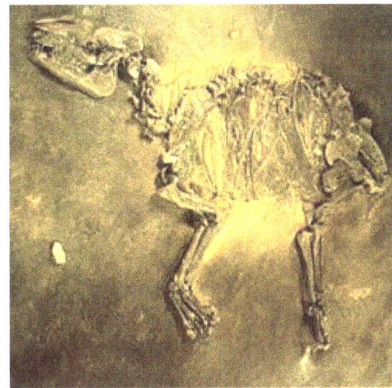

Dinosaur   Mammal

In the fossil record dinosaur fossils are clearly identified as dinosaurs and mammal fossils are clearly identified as mammals. But fossils depicting the long evolutionary history between the two groups are totally missing.

Mammal to man

Man has been classified by secular biologists as belonging to the order Primates. The order Primates has supposedly evolved from an insectivorous ancestor. Are there fossil remains that would document the long history of this evolution? Elwyn Simons, a leading authority in the field of primates has stated, "In spite of recent

[23] Werner, Carl. *Evolution: The Grand Experiment*. New Leaf, 2007. 126. Print.
[24] Simpson, G.G. *Tempo and Mode in Evolution*. New York: Columbia UP, 1944. 105. Print.

finds, the time and place of origin of the order Primates remains shrouded in mystery."[25]

Another leading authority in paleoanthropology, A.J. Kelso commented, "The transition from insectivore to primate is not documented by fossils. The basis of knowledge about the transition is by inference from living forms."[26]

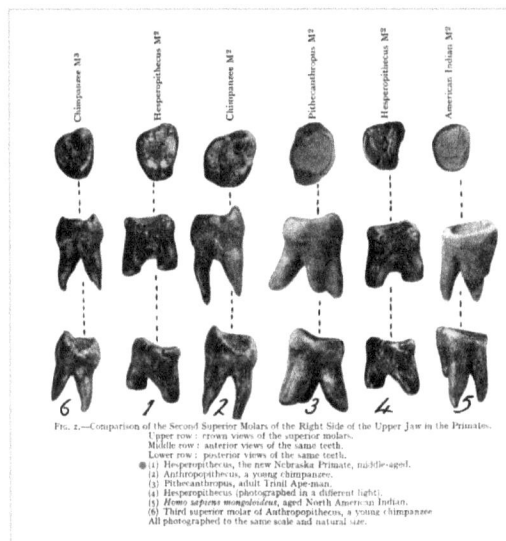

Despite the well-constructed museum displays depicting the evolutionary history between man and an ape-like creature, the evidence is scant. The artists' renditions are often based on imagination constructed from a few bits and pieces of bone and teeth. The first picture above is of the so-called Nebraska Man built up from what was later identified as a single tooth of an extinct pig.

[25] E.L. Simons, New York Academy of Science 167:319, 1969.

[26] Kelso, A.J. *Physical Anthropology*. 2nd ed. New York: J.B. Lippincott, New York, 1974. 142. Print.

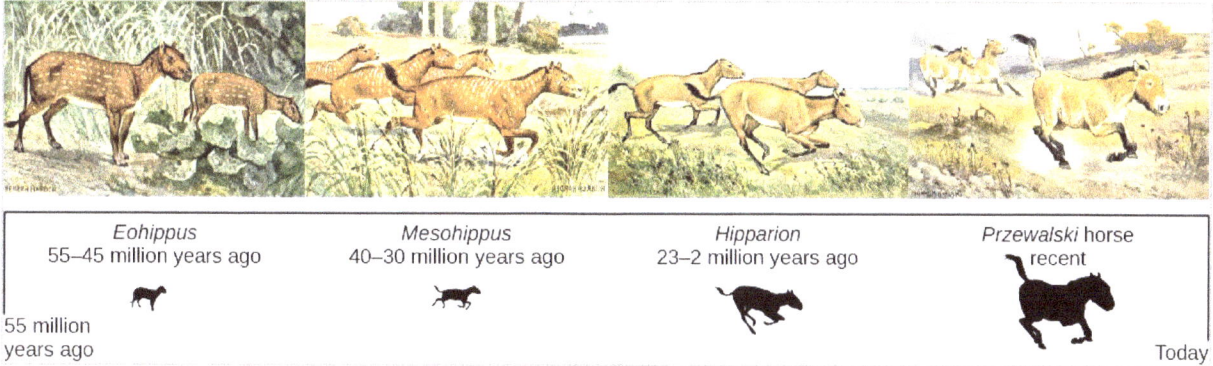

| Eohippus | Mesohippus | Hipparion | Przewalski horse |
| 55–45 million years ago | 40–30 million years ago | 23–2 million years ago | recent |

55 million
years ago

Today

The Horse Series as it was called was one of the show cases for evolution for many years. The various horses are lined up as if they evolved through time to become our modern horse. Certainly this would show that mammals had evolved in addition to the dinosaurs. The picture above is a standard picture used in many textbooks and resources.

For many years book publishers published the above chart, or ones like it, and it continues to be published in many books and portrayed in museum displays today. Here is an example of what our children are learning. The popular coloring book, *The Evolution of the Horse* says:

> The horse family tree branched many times during its 55-million-year history. Of the many scores of horse genera, from the small Eocene Hyracotherium (the old Eohippus) to Hipparion horses at the end of the last Ice Age, only one genus, Equus remains.[27]

---

[27]Wynne, Patricia J. *The Evolution of the Horse*. Minneola, NY: Dover Publications. Print.

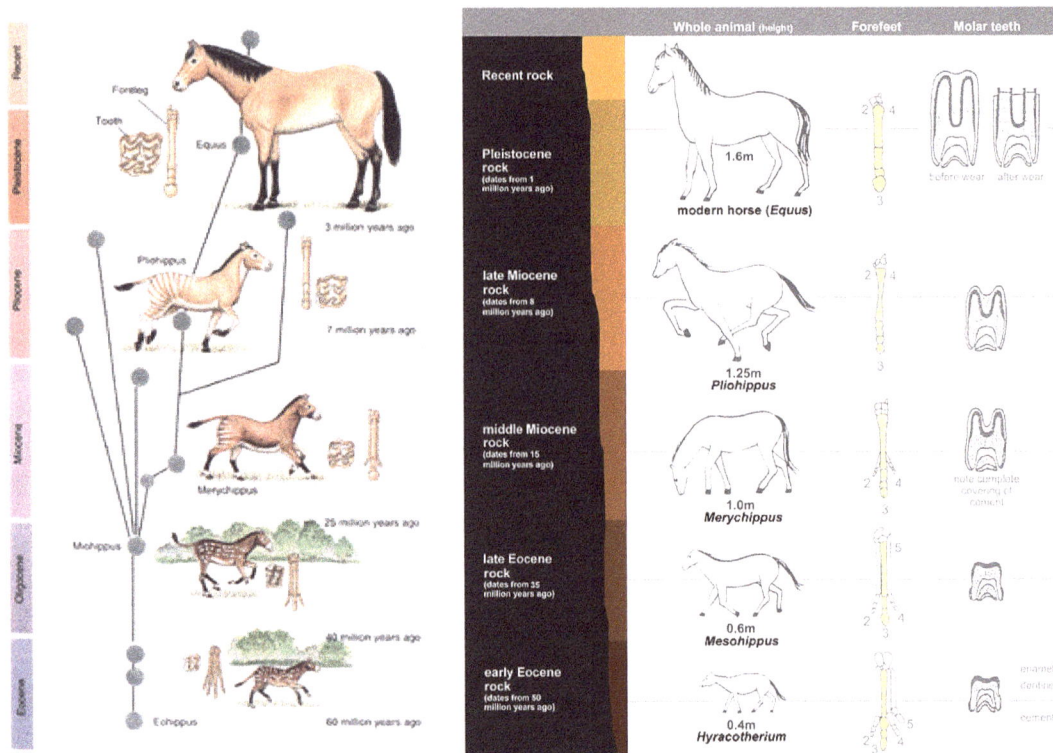

There are many depictions of the Horse Series.

The popular Wikipedia online encyclopedia even today puts it this way:

> The evolution of the horse occurred over a period of 50 million years, transforming the small, dog-sized, forest-dwelling *Eohippus* into the modern horse. Paleozoologists have been able to piece together a more complete outline of the modern horse's evolutionary lineage than that of any other animal.[28]

Just how scientific is the horse series evolution icon? Consider these points.

- The first animal in the charts has traditionally been called, Eohippus (dawn horse.) When this animal was discovered in Europe, it was classified as Hyracotherium, because it was thought to resemble a hyrax. When the same creature was discovered in North America, it was named Eohippus. But it really has no connection to the horse at all and was reclassified as Hyracotherium. Many textbooks continue to show this animal as Eohippus.
- The horse series was constructed from fossils found in many different parts of the world, and nowhere does this succession occur in one location. The series is

---

28 Web. 9 Jan. 2015. <http://en.wikipedia.org/wiki/Evolution_of_the_horse>.

formulated on the assumption of evolutionary progression, and then used to supposedly prove evolution!

- The number of ribs varies within the series, up and down, between 15, 19, and 18. The number of lumbar vertebrae also changes from six to eight and then back to six.
- There is no consensus on horse ancestry among paleontologists, and more than a dozen different family trees have been proposed, indicating that the whole thing is only guesswork. Fossils of the three-toed (Mesohippus) and one-toed species are preserved in the same rock formation in Nebraska USA, proving that both lived at the same time, strongly suggesting that one did not evolve into the other. The misnamed *horse series fossils* rather show variation within the Horse Kind who all lived at roughly the same time and over wide geographic areas.

Read the following statements by some of the paleontologists who were asked specifically about horse evolution as presented in the previous charts:

> At present, however, it is a matter of faith that the textbook pictures are true or even that they are the best representations of the truth that are available to us at the present time. ... It makes quite a difference whether a name on a diagram represents a whole skeleton or just a tooth....[29]

Kerkut's main problem with the horse series is that the original fossils are not available - everything on display is a reproduction, and there's no way of knowing which bones were really found and which were added from imagination. He refers to the common practice of *reconstructions* in textbooks and museum displays, where a full image of a presumed ancient creature is based on just a few actual fossil bones.

From Dr. Niles Eldredge:

> I admit that an awful lot of that has gotten into the textbooks as though it were true. For instance, the most famous example still on exhibit downstairs (in the American Museum) is the exhibit on horse evolution prepared perhaps 50 years ago. That has been presented as literal truth in textbook after textbook. Now I think that that is lamentable, particularly because the people who propose these kinds of stories themselves may be aware of the

---

[29] Kerkut, G.A. *The Implications of Evolution.* New York: Pergamon, 1960. 141-149. Print.

speculative nature of some of the stuff. But by the time it filters down to the textbooks, we've got science as truth and we've got a problem.[30]

From D.M. Raup:

The record of evolution is still surprisingly jerky and, ironically, we have even fewer examples of evolutionary transition than we had in Darwin's time. By this I mean that the classic cases of Darwinian change in the fossil record, such as the evolution of the horse in North America, have had to be modified or discarded as a result of more detailed information—what appeared to be a nice simple progression when relatively few data were available now appears to be much more complex and less gradualistic.[31]

And finally, from Dr. John Morris:

It is now acknowledged that horse evolution as recorded in the fossils follows no recognizable pattern, and that the evolutionary "tree" looks more like a multi-branching "bush." The successive forms indicating straight-line evolution appear only in textbooks; they do not appear in the fossils. Sometimes fossils of different types that supposedly lived at different times appear together in the same strata layer. In Oregon, the three-toed grazer Neohipparion (very much like Merychippus) has been found with Pliohippus. In the Great Basin area, Pliohippus has been found with the three-toed Hipparion throughout the timeframe supposedly represented. Evolutionary scientists freely admit this situation--and to their credit often attempt to correct the misconceptions--but still the horse series appears in the textbooks.[32]

Horse evolution is supposed to represent the textbook example of transitional evolution. These types of illustrations deeply affected me when I was a kid. But if this is the best example, then evolution seriously lacks the scientific support that it desperately needs. The fossil evidence rather supports the horse kind with variations within the kind.

---

[30] Dr. Niles Eldredge, curator at the American Museum of Natural History, in a recorded interview with Luther Sunderland, published in *Darwin's Enigma: Fossils and Other Problems*. Master Books: El Cajon, California, USA.

[31] Raup, D.M. "Conflicts between Darwin and Paleontology." *Field Museum of Natural History Bulletin* 50 (1979): 22. Print.

[32] Morris, John. "The Mythical Horse Series." *Acts & Facts* 37.9 (2008): 13. Print.

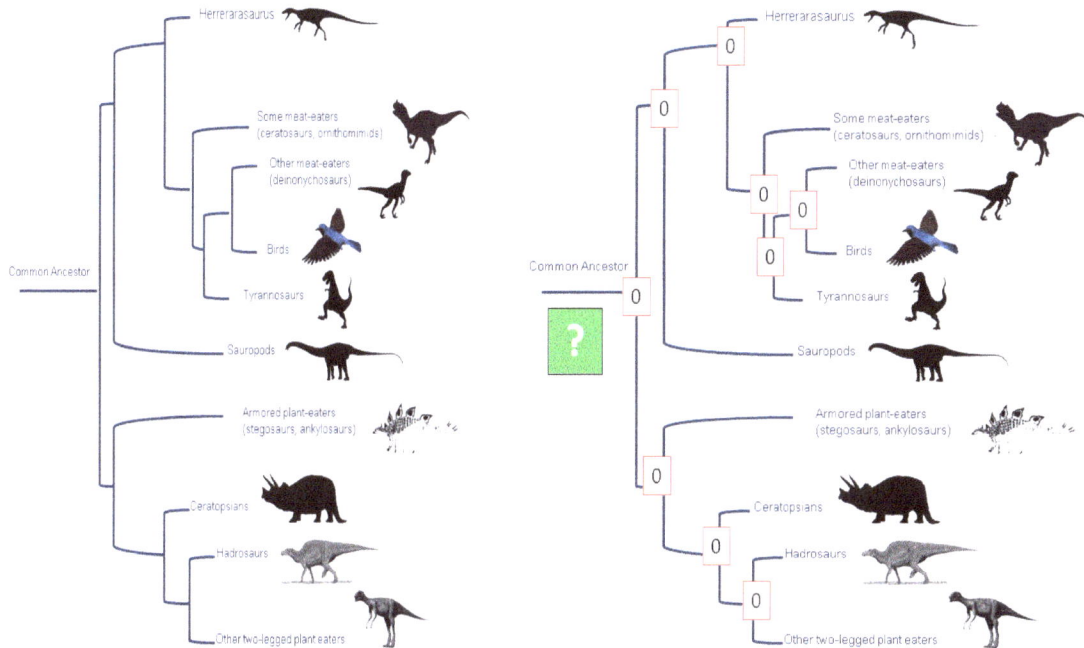

The chart on the left is based on a display in the Chicago Field Museum. This one clearly implies that the links among the ten different kinds of dinosaurs along with the link to the common ancestor of dinosaurs is a fact of science. The reconfigured chart on the right shows the number of fossils that have been found to support this idea. The evidence clearly shows that there are no transitional fossil forms connecting the various kinds of dinosaurs between the various kinds, to the various kinds and to the supposed common ancestor. Is this scientific or a product of biased and wishful thinking?

Activity 8

Using internet or some other resources, compile a list of transitional fossils that have been accepted and promoted by evolutionists. What kinds of questions might you ask about these various transitional fossils?

Please take Quiz #8, Appendix B

Lesson 9 – The Two Contrasting Views of Man

The Influence of the Enlightenment

The Age of Enlightenment; also called The Age of Reason and The Age of Revolution - Mainly covering the 18ᵗʰ and 19ᵗʰ Centuries, it was a time of questioning the "old ways" of the church, the Scriptures and the Biblical view of man. Our modern age is a direct consequence of the thinking that sprouted during The Enlightenment. These two pictures depict the spirit of the Age of Enlightenment, as realized in the French Revolution. Order and compliance during the time of the Revolution was assured by the frequent use of the guillotine.

Words to know: *The Enlightenment superstition cognitive verify scrutiny scientific method myth transformation deceive enthrone absolute suppress autonomy The Renaissance*

The Enlightenment View of Man

The age known as **The Enlightenment** (1700s-1800s) brought about a change in the way man viewed himself, nature, God and the world around him:

1. Man began to shun the **superstitions** of the previous age. This would also include the Bible which was being viewed more suspiciously than in the past.
2. Man increasingly viewed God as distant and uninvolved in the world.
3. Man increasingly trusted his own **cognitive** abilities to discover, to define and to **verify** truth.

4. In short man increasingly saw himself as more in control of his world and God as less in control. This would include his ability to change the world through education and social change or social engineering.

The Enlightenment really had its beginnings back in the **Renaissance** (meaning *rebirth*), when anything ancient was idolized. This period covered primarily the 13th through the 15th centuries. During this time the Bible began to be increasingly viewed with suspicion and doubt. After all, it was thought that the ancients held the key to discerning man's historical past.

It was during this same period of time, however, that ancient Greek, the language of the New Testament was rediscovered. Many translations of the Bible came out of this time including the first English translations. The event known as The Reformation was a direct result of this rediscovery of ancient Greek. Many historians credit the origins of science as a discipline of exploration and discovery to these changes. In the 1600s science had enjoyed a brief period of discovery and boom based on the fact that God had created the world and therefore it could be organized and categorized according to laws He had established.

But another movement was also in progress – The Enlightenment. Man transitioned during The Enlightenment. Modern science was born. Truth would henceforth be subject to man's scientific judgment and **scrutiny**. By the 20th Century this judgment and scrutiny would apply to all things, not just the world of nature. This produced a kind of social engineering. Through man's education and control, the world could become a utopia, a heaven on Earth. (Isn't it interesting that these ideas gave birth to Nazism and Communism.) Science as it had been practiced became a type of belief cloaked as science.

We usually define science as knowledge. In practice, modern science consists of values and norms defined by groups of academic science scholars, including biologists, paleontologists, geologists, psychologists and even medical doctors. Instead of discovering nature's God through exploration and investigation, as early naturalists had done, man dethroned nature's God and enthroned himself as God. God, nature, and facts would from now on be subject to scientists – men academically trained in the **scientific method** who would rely on experimentation and observation *alone* to prove truth. At least that was the stated noble goal. Stories like the creation, the flood of Noah and miracles were beyond scientific discovery and so were out. Man was now enlightened, and no enlightened man would believe in things like *myths* and scientifically improvable stories. The whole culture of the 1700-1800s underwent a complete *transformation*. This gave us the modern age where belief, cloaked as science, now sits in control. This "science" now decides what will and what will not be taught in our public schools and has increasingly more influence in public policy and practices.

Isaac Newton (1642-1726), Kepler (1571-1630) and Galileo (1564-1642) – all brilliant scientists who believed in the recent creation of *Genesis* and a global flood. All this changed with the advent of The Enlightenment view of creation and man. Science increasingly moved away from a Biblical explanation of origins and earth history.

All this seems like the reasonable thing – to prove things by scientific method and hard investigative facts. After all, we do not live by myths and *just so* stories. But man has forgotten who he is. He thinks he has the power and capability to discern truth apart from any sort of revelation by a God he cannot test and verify. In reality, man has *deceived* himself. He has developed a worldview that excludes the God of the Bible. He has *enthroned* himself as the god of nature. Does man really have this kind of power and sufficiency of knowledge that he can decide if there is or isn't a Supreme Being? Does man really have the moral absoluteness and perfection to define values and morals? Two World Wars and hundreds of millions of people destroyed in the process have sufficiently demonstrated that man is incapable of defining moral values. It is time to get back to a standard that has stood the test of time and change, and has been demonstrated to be the one absolute standard revealed to us by the God who created us and our Earth.

Does the Bible adequately explain man's nature? Does its view of man match up with what we have observed through the ages? More importantly, if the Bible's view of man is correct, how would it affect his rational and behavior in the field of the sciences? Would it have any effect?

The Biblical View of Man

The Bible tells us in the first three chapters of Genesis that man was originally created without sin and moral fault. Then out of rebellion against his creator, he ruined himself, his relationships and the natural world into which God originally placed him. Sin and moral imperfection were now part of his nature and he would from then on pass on this ruin to his descendants. Genesis chapter three succinctly describes the consequences of this rebellion:

1. He was afraid of God
2. He hid himself from God's presence and moral light
3. He passed blame onto his wife
4. He succumbed to moral temptation
5. He brought about enmity and distrust
6. He became susceptible to deceit and being deceived
7. He opened the door of the world into which God placed him to God's enemies

Adam and Eve Driven out of Eden – Gustave Dore (1832-1883); one of the classic pieces of artwork depicting the Genesis event known as The Fall. Having abandoned this piece of Genesis history, man has exchanged a totally naturalistic view of man and of the world around him. Most people do not see the connection between this new evolutionary view of man and the chaos in our world today.

The Bible's recorded history of man from that point is replete with examples of how these things were born out. One only needs to study the life of Jacob to see these consequences come to pass again and again. Jeremiah put it this way when prophesying to Judah about God's judgment, "The heart of man is more deceitful than all else and is desperately sick; who can understand it?" (Jeremiah 17:9). Paul later in the New Testament expressed it this way, "(Man) **suppresses** the truth in unrighteousness." (Romans 1:18). In other words, like Adam, man has his own agenda and ideas and they don't line up with God's. Therefore, he does what he can, deceitfully, to put down and hide the truth of God and his own accountability. How would this affect man's view of his world? He has two choices. Either he agrees with what God has said about himself and submits or he tries to cover it up and seek alternative explanations and excuses to remain in control of his life. He will seek to rule his own world and make up his own laws to comfortably govern his own life. How do you think this would affect his judgment when it comes to areas in science? With things that are morally neutral like electric light bulbs, man will reflect his creative abilities given to him by God. Man's **autonomy** is not threatened by these morally neutral scientific accomplishments. With things that have moral implications like abortion, he will argue for a scientific explanation of human life and against what the Bible has to say because they hold man accountable to some One higher in authority. How about things like the Big Bang which is man's explanation of the origin of the universe? If man is subject to the God who created him, he will look to His revelation of Scripture for guidance. If he values his own abilities to arrive at truth over those of God's Scripture, he will seek to explain away that revelation. This is the nature of man whether he is a Christian or not.

Psalm 53:1-4 summarizes an indictment against man. Romans chapter 3 repeats this indictment:
1. No man understands the truth apart from God's ways
2. No man seeks after God
3. Every man has turned aside
4. Together man has become corrupt
5. Mankind is called "the workers of wickedness"
6. They persecute God's people and do not call upon God

The Bible paints an unflattering picture of man. Psalm 53 tells us how he approaches life. It is who he is. And yet in the church many Christians are willing to let man, "the scientist," interpret the Scriptures by what he says that he has discovered. If man is corrupt and does not understand the truth on his own, why would people, especially those who are followers of God, depend upon man to sort out issues in Genesis like the age of the Earth and Biblical chronology? Does somehow being a scientist make things different? Does it qualify a person to interpret God's words? Rather, it would seem that the person best able to interpret God's words is the one who does just the opposite of the above. The person who seeks God and His ways would have understanding. The person who would turn aside from his own ways would have understanding. The person who would honor God's people would possess proper perspective. In fact Jesus demonstrated this very thing in Matthew 4:4 where He stated, "Man shall not live on bread alone but on every word that proceeds out of the mouth of God."

The Effect on Education

Why would man honor academic knowledge above the word of God? A quick reading of the New Testament Book to the Roman church will state the reason succinctly – verse 18 of chapter one tells us that the history of man is filled with examples of choosing what he wants rather than what God wants. This self-centered behavior has not only spoiled the creation, but has shrouded God's plan for man in confusion. On the outside an academic education seems like the normal thing to do in one's life. On the inside of many of our schools, however, an alternate view of man, God and the world are being inculcated into the minds and hearts of young people. In the name of academic and scientific integrity the Biblical explanation of man's purpose and meaning has been ridiculed and mocked so that all man is left with is an evolutionary explanation which leads to despair and an immoral culture. Throughout this ordeal it is claimed by the scientific establishment that there is abundant fossil evidence to establish the evolutionary history of man. In our next lesson we will explore this fossil evidence and see if it is as clear as is taught.

Activity 9

Find three or four papers, articles or books about The Enlightenment. What kinds of things did you notice about the view of man, politics, science, and God during this time?

Please take Quiz #9, Appendix B

Lesson 10 – The Fossil Evidence and the Evolution of Man

Ernst Haeckel, about 1905, Germany's popularizer of Darwin's ideas in the 19ᵗʰ Century. In 1865 Haeckel stated, *"If we do not accept the hypothesis of spontaneous generation (life from non-living matter), then at this one point of the history of development (evolution) we must have recourse to the miracle of a supernatural creation."*[33]

Words to know: anthropology Pandora's Box preconceived homology pathologist mineralized dispersion authenticated Pliocene circular reasoning hominid

The glory of man

Perhaps the most difficult area for evolution lies in the discipline of **anthropology**. More false starts, frauds and vacillations have occurred in assigning fossils to various stages of the evolution of man than in any other area

---

[33] Taylor, Ian. *In the Minds of Men*. Toronto, Canada: TFE, 1987. Print.

of paleontology. Modern evolutionists have ascribed this to the scant remains that have been discovered. That may be the case. But the lack of fossils should then tell us that when it comes to the evolutionary history of man, caution would be the most reasonable approach.

Underneath the search for the evolutionary history of man lies an erroneous belief about man himself. If man is as old as evolutionists claim he is, then there should be no lack of fossil remains to discover and explore. According to population statistics, the present population of the world today can be arrived at in a short few thousand years starting with just two people. But man is supposed to be magnitudes older than just a few thousand years. Where are all the graves? Could it be that the Bible is historically accurate in its record and that man was created on this planet just 6,000 years ago?

The Bible proclaims in the first chapter of Genesis that man was created in the image of God – male and female. What gives man a sense of importance and value is this fact. Without it, man is just another animal with a false sense of importance. *Made in the image of God* explains a lot about man's uniqueness in this world and indeed in all of history. This is his glory! John 3:16 makes perhaps the clearest statement about the Biblical view of man. God so loved man that He gave His one and only Son to save him from his sin. Nothing in science can explain this position man holds with God – nothing! I believe that this is the main reason paleontologists have made virtually no progress in advancing their belief that man evolved from ape-like creatures. There is no fossil evidence to support it. I have followed this adventure since the 1970s, reading article after article that have reported the latest fossil finds. Every one of the examples in them has been either reclassified or discarded for one reason or another. The only reason for this is that man persists in his rebellion against his Maker and continues to cling to the false hope that the next fossil link is just around the corner. Let's now take a look at some of the fossil evidence for the evolution of man.

Without the miracle of the direct act of the Creator God, man would not exist. He is indeed a unique creation. So then, what is all the fossil evidence about? What do we make of human-like skulls and bones thought to be man's ancestors, but which do not look much like the man of Genesis chapter one?

Stammbaum des Menschen.

Menschen

Gorilla — Orang
Schimpanse — Gibbon
Anthropoiden — Fledermäuse
Hufthiere — Affen — Nagethiere
Faulthiere — Raubthiere
Walfische — Halbaffen
Beutelthiere
Ursäuger (Promammalia) — Schnabelthiere

Säugethiere (Mammalia)

Knochenfische (Teleostei)
Molchfische (Protoptera)
Vögel (Aves)
Schmelzfische (Ganoides) — Reptilien — Schildkröten
Amphibien — Crocodile
Lurchfische (Dipneusta) — Eidechsen
Petromyzon — Urfische (Selachii) — Schlangen
Myxine — Kieferlose (Cyclostoma)
Schaedellose (Acrania) — Amphioxus

Wirbelthiere (Vertebrata)

Insecten — Ascidien
Crustaceen — Chordonier — Salpen
Gliederthiere (Arthropoda) — Mantelthiere (Tunicata)
Weichwürmer (Scolecida) — Weichthiere (Mollusca)
Sternthiere (Echinoderma) — Ringelwürmer (Annelida)
Nesselthiere (Acalephae) — Urwürmer (Archelminthes)
Pflanzenthiere (Zoophyta) — Würmer (Vermes)
Schwämme (Spongiae)
Gastraeaden.
Eithiere (Ovularia) — Planaeaden — Infusionsthiere (Infusoria)
Synamoebien
Amoeben
Moneren

Wirbellose Darmthiere (Metazoa evertebrata)

Urthiere (Protozoa)

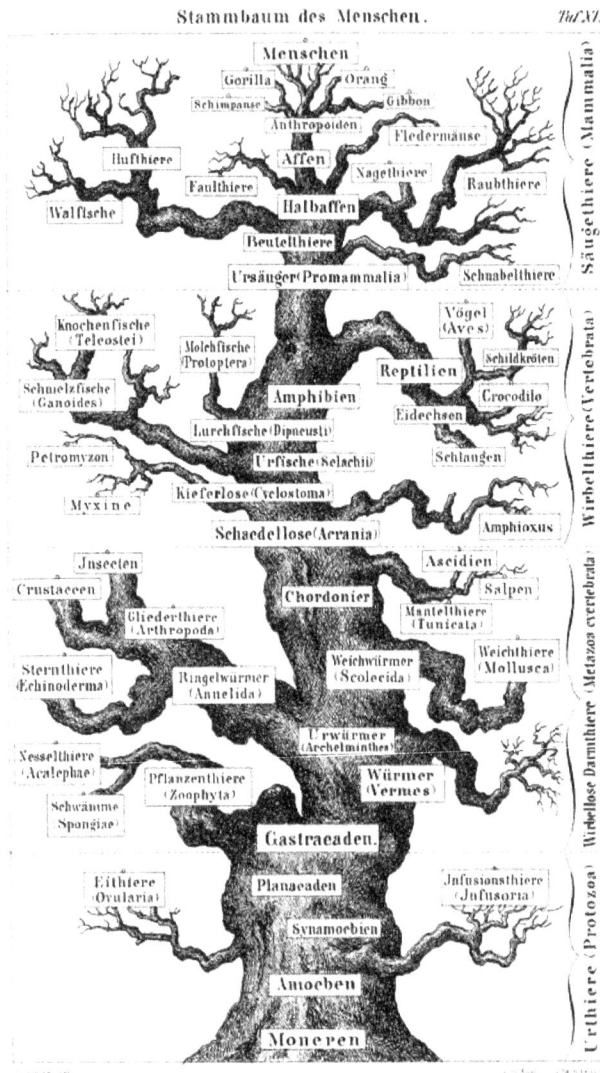

> This is Man's family tree according to Ernst Haeckel (1874). Many paleontologists now reluctantly admit that there is not a shred of evidence for the trunk or the main branches. It is all a very imaginative drawing for which Haeckel was famous.

When Darwin's book, *On the Origin of Species by Means of Natural Selection, Or The Preservation of Favored Races in the Struggle for Life* was published in 1859, he left out any references concerning the application to the evolution of mankind. The culture into which Darwin's book was introduced was still very much staunchly convinced that *Genesis* was true. But Darwin's book opened up a **Pandora's Box** and many understood its meaning as it applied to mankind. The search was on for the missing links in man's evolution. By 1871, when Darwin published *Descent of Man*, opposition to Darwin's ideas had pretty much subsided.

One of the most blinding facts of anthropology has been that it has been based on *looking* for the evidence to support a **preconceived** idea rather than formulating conclusions based on evidence. One of the first discoveries to spark controversy was the discovery of Neanderthal Man.

Neanderthal Man

In 1856 a skull cap and some limb bones were found in the Neander Valley in Germany. Just over a year earlier the great anatomist, Sir Richard Owen, the inventor of the word, *dinosauria*, had addressed the Royal Institution of Great Britain on the significant differences between ape and man which he claimed precluded man's link with the ape. The remains found in the Neander Valley, however, generated lots of controversy. What kind of a creature did these bones belong to? This was three years before Darwin's *Origin* and fifteen years before his book, *The Descent of Man* were released. Still, the imaginations had been stirred by geologists of the early 1800s. Their insistence that the Earth was older than what the Bible said and that living things had obviously changed over the millennia had already provided the framework. Could these remains from the Neander Valley belong to some ancient ancestor of man's?

Thomas Huxley (Lesson 7) knew of the Neander Valley remains but considered them human. He had already been convinced that man was no different than the brutish apes, at least in structure. He used a comparison chart of the skeletons of various apes and man to prove that there was no inherent difference between man and the apes.

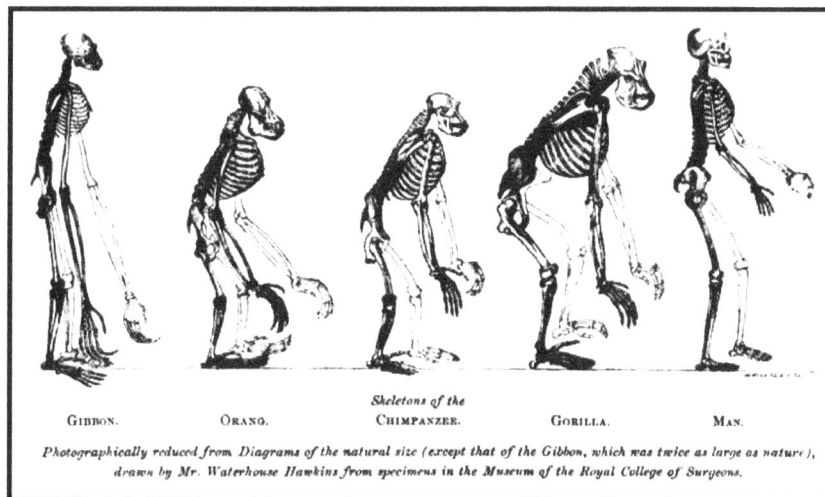

GIBBON.   ORANG.   *Skeletons of the* CHIMPANZEE.   GORILLA.   MAN.

Photographically reduced from Diagrams of the natural size (except that of the Gibbon, which was twice as large as nature), drawn by Mr. Waterhouse Hawkins from specimens in the Museum of the Royal College of Surgeons.

Thomas Huxley's skeleton comparisons of anatomical structures (as drawn by Waterhouse Hawkins) – *homology*, the study of similar outward characteristics, has dominated evolutionary biology for many years. Its modern version is called cladistics. The main difference is that paleontologists of today do not place the same importance on the fossil evidence that Darwin originally thought it should. If a creature shares similar characteristics, then it is related, regardless of where fossils might fit into the scheme.

The idea of an ape-like ancestor of man did not originate with Darwin. It was a natural conclusion of the Deistic thinking that the natural world was governed by natural law. The idea that man had developed from former ape-like ancestors, therefore, was not totally strange. Deism proclaimed that no God was involved in the development of man. Huxley wrote,

> But if Man be separated by no greater structural barrier from the brutes than they are from one another, then, there would be no rational ground for doubting that man might have originated...by the gradual modification of a man-like ape or as a ramification of the same primitive stock as those apes...Man is, in substance and in structure, one with the brutes. [34]

Bear in mind that there were no fossil remains of any of the so-called ancestors of man at this time. These conclusions were actually just assumptions based on a bias against the Christianity of the day. Future "evidence" would simply be pasted into the pattern that had already been determined about man's origin.

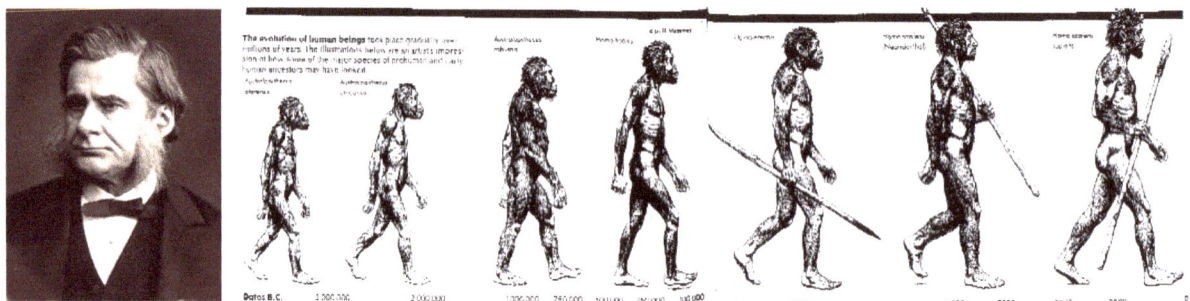

Thomas Huxley, *Darwin's Bulldog*, and the logical conclusions that followed what he advocated.

Many **pathologists** of the day had a chance to examine the Neander Valley finds and their insistence was that this creature *appeared different* than Homo sapiens because of *deforming diseases* present during his life. Over the next few decades parts of over 60 individuals were found mostly in Europe, though some had been found in Africa and in Asia. The pathological evidence, however, was ignored because the fever to discover man's ancient evolutionary remains overshadowed any science of the day. A few decades later French paleontologist Marcellin Boule declared that Neanderthal was *not* human and not even ancestral to humans. Boule regarded him

---

[34] Huxley, Thomas A. *Evidence as to Man's Place in Nature.* 1863. Print.

as an extinct side branch of the evolutionary tree. The Neanderthal remains had previously been classified as Homo antiquus (*ancient man*). They were subsequently changed to Homo neanderthalensis. The popular press of the day called the remains, Neanderthal Man and this became fixed in the public mind from that time on as a *missing link*. I was taught as a young boy that Neanderthal Man was one of the missing links between apelike creatures and man. Today Neanderthal has been reclassified again to reflect the original pathological findings. He is no missing link at all, but man with debilitating diseases acquired from some unknown event or source. No one really knows. Some of the diseases linked to Neanderthal, however, seemed to be related to a lack of sunlight and proper diet. Could these have been related to the effects of the Ice Age after the Flood?

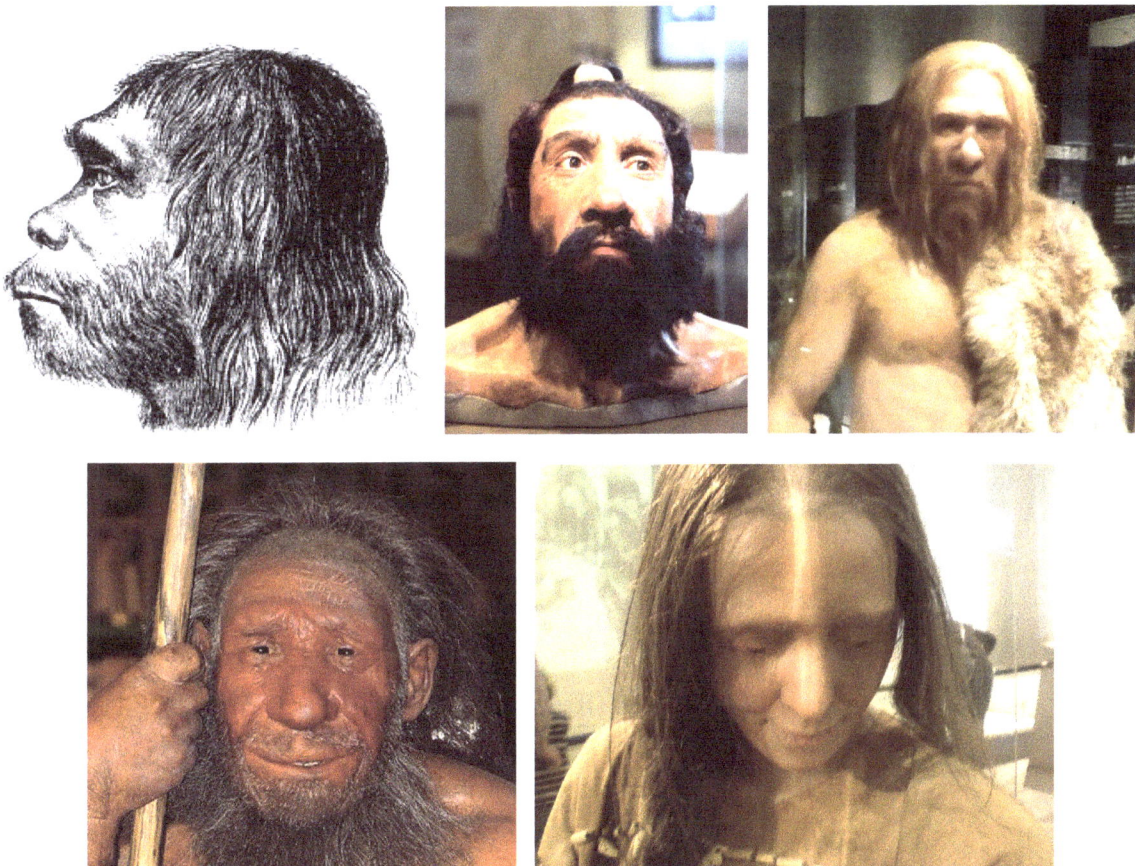

The changing face of Neanderthal Man. The pictures reflect the biases of presenters over the years. Some have commented that if he was dressed in a suit and shaved, he would look every bit the business man of downtown New York.

Dr. John Morris comments,

Far from being an evolutionary ancestor or even a sister group among the primates, Neanderthals were fully human. Perhaps they were a family group that migrated from

Babel after the confusion of tongues. They lived primarily in Ice Age Europe and developed a primitive lifestyle and physical adaptations to cope with the cold. They employed agriculture, used weaponry, practiced religious rites, buried their dead, adorned their bodies, and played musical instruments—all aspects of human culture.[35]

The Calaveras skull

In 1866, 130 feet below ground in the gold-bearing gravels of the Sierra Nevada, California a skull was discovered that was almost entirely *mineralized*. The skull was *authenticated* by J.D. Whitney himself, chief of the California Geological Survey at the time. It had been found in what had been identified as *Pliocene* strata, five million to two million years old. There was one small problem, however: the skull was too modern to be dated Pliocene. Also, along with the skull were found stone mortars. This piece of paleontology was an embarrassment, as it was firmly believed that during the Pliocene time man was in the infant stages of his evolution. This skull was modern! The skull was declared to be a hoax and quietly set aside along with Whitney's detailed report. By the way Mt. Whitney in California is named after this great geologist.

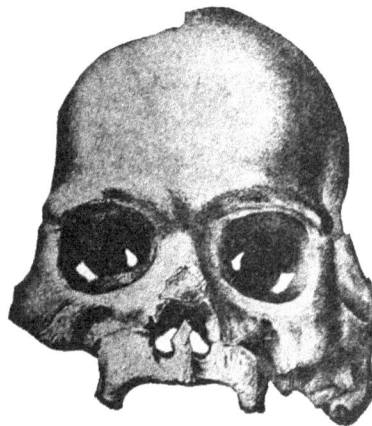

The Calaveras Skull – modern in every detail, was found alongside of some stone mortars which are not mentioned when the skull is discussed today.

Cro-Magnon man

A number of skeletons were found in a cave in France in 1868. Since then, more skeletons have been found in other parts of Europe. The Cro-Magnon, as they were

---

[35] Morris, John, and Frank K. Sherman. *The Fossil Record*. Dallas: Institute for Creation Research, 2010. 77. Print.

called, were truly human, some being over six feet tall and having a talent for cave art. It is the Cro-Magnon more than any other who has documented his times in beautiful cave art, even depicting woolly mammoths! The description of *Cave Man* continues to be used when describing this man, as if he was some brutish evolutionary link. Was he an evolutionary missing link, or a product of the **dispersion** from the Tower of Babel? When I was a kid, I was taught that he was another missing link in the evolution of man.

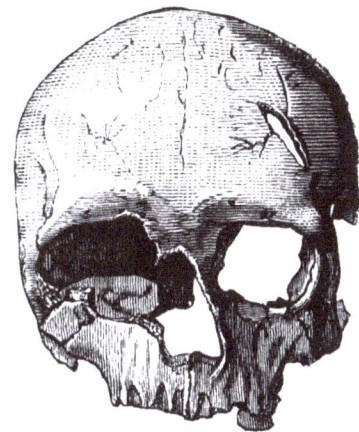

The idea that Cave Men represent an evolution from brute to modern man is unfortunate. If a Biblical view is accepted, then the harsh existence of man following the Babel dispersement would make sense. Some men could have been technologically inclined, while others, such as Cro-Magnon may have lacked technical skills for building, but were more artistically inclined. These might have gravitated to caves, where we see their paintings today

Java Man

In 1877 Eugene Dubois entered Jena University under the tutelage of Ernst Haeckel. Haeckel fired the imagination of this young man of 19 to the point where Dubois decided he was going to set out to find the missing link between man and ape. Over eight years a few fossils were found in different locations in Java. As Dubois was not actually present when a couple of the fossils were found, their exact location with respect to each other varies from one report to another. Of course, the interpretation of these bones would not be reliable and would be subject to the bias of the one in control of the project.

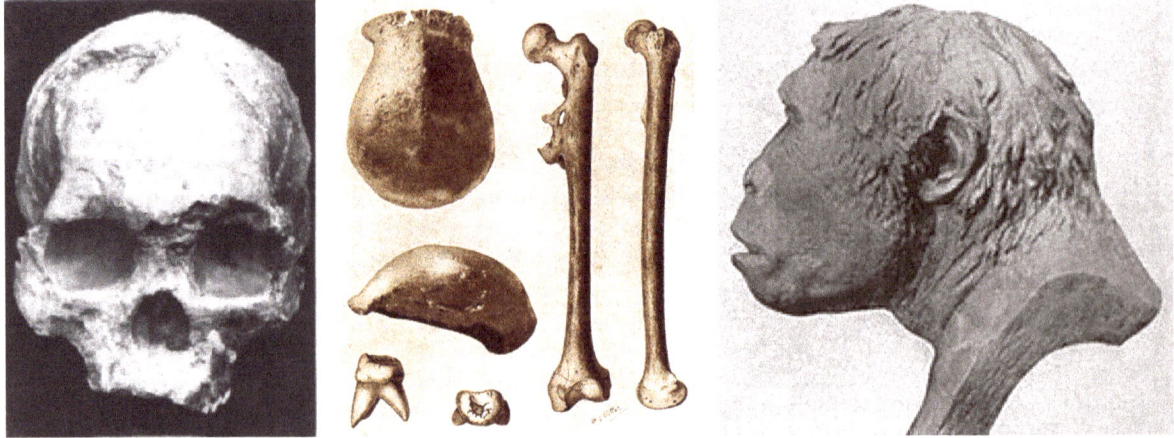

The first picture is of a fossil human skull given to Dubois in 1889 in Java in the same strata from which another skull in Australia was found and dated at 500,000 years old! The skull is perfectly human. The second picture shows the fossils Dubois found over the next eight years, spread out over different locations. The third picture is a reconstruction of what little evidence there was! The picture, however, conveys a thousand words!

5 Nisan 1964 tarihli
Sunday Times'da yer alan çizim.

Maurice Wilson'un
çizimi.

N. Parker'ın çizimi.
N.Geographic, Eylül 1960

The different artists' renditions that have helped to shape public perceptions of *Java Man*.

Dubois immediately dubbed one of the thigh bones as Anthropopithecus erectus, meaning *upright, man-like ape*. After some further thought he recognized the evidence was scanty and the thigh bone was perfectly human-looking. He renamed it Pithecanthropus erectus, meaning *upright ape man*. This conclusion was reached in

1893 and this is what Dubois held to for the rest of his life until his death in 1940. Subsequent investigations revealed that one of the femurs and the skull cap had come from *different individuals*. Dubois was convinced that they were from the same individual and this is what drove his conclusion that he had found the missing link. Then in 1950, ten years after his death, Dubois' Pithecanthropus erectus was reclassified and renamed Homo erectus, meaning *upright man*. Yet this Java Man continues to find its way into textbooks and museums as authentic evidence for man's evolutionary ancestry. The whole story demonstrates how biases can blind a person from seeing his own **circular reasoning**.

Piltdown Man

The most amazing fraud of the 20th Century has to be Piltdown Man. Although it is a mystery as to who actually perpetuated this fraud, many think it was the famous author of the Sherlock Holmes stories, Sir Arthur Conan Doyle! Sir Arthur was a spiritist – one who believes in and communicates with spirits. The scientific community of the day, of course, was skeptical of his spiritism. Deistic ideas had shaped the thinking that the spiritual world was myth. They spared no effort in criticizing Sir Arthur's beliefs. Many think that the Piltdown hoax was Sir Arthur's way of getting back at these scientists.

Below is a picture of the skull that was discovered in the gravels of Piltdown, England. Alongside it is an artist's rendition of Piltdown Man that has led millions of people astray with the notion that he was one of the missing links in the evolutionary history of man.

Skull that was discovered in 1912 and artists' renditions of Piltdown Man.

The skull was originally discovered in 1912 in England during the fever-pitch years in the search for man's evolutionary ancestors. For over 40 years this find fooled the scientific community and even today shows up in some textbooks as an example of a missing link. In the 1950s after modern updated chemical tests on the skull, it was discovered that the skull was really a clever compilation of human and orangutan skull parts. The teeth had been filed down to give them age and the skull had been treated with fluoride to make it appear weathered. This is the kind of "science," however, that has filled the evolutionary search for man's missing links.

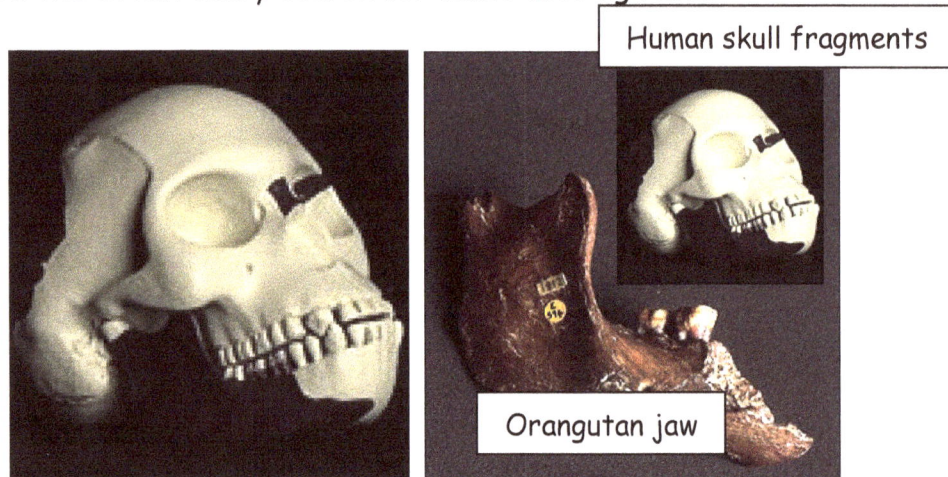

This is a representation of Piltdown Man – a very clever hoax that fooled the science community (and millions of people) for over 40 years! It appears that evolutionists saw what they had expected to see – a *transitional* link.

Even though most modern textbooks have expunged Piltdown Man from man's evolutionary tree, he is nevertheless an important part of demonstrating the bias of man in science. Most people mistakenly believe that scientists are unbiased and fair in their scientific quests. Every human being has biases. We may not be willing to admit it, but it is part of our fallen nature. Most scientists when confronted with this fraud simply comment that this is the self-correcting nature of science. This totally misses the point. The quest to discover man's evolutionary links has been characterized by this kind of biased "science." It betrays a bias, not scientific investigation.

Rhodesia Man

Accidentally discovered in 1921 by zinc miners in Zambia (formerly British Rhodesia), the skull was almost complete and had the appearance of being ancient. It was initially proclaimed a missing link and named Cyphanthropus or Stooping Man. Debate ensued and subsequently the skull was renamed Homo rhodesiensis and recognized as a true man. The initial discovery, however, is what sticks in the minds of people. The bone was heavily mineralized and presumed to be very old.

One curious thing about this skull, however, that managed to escape attention for some time was the existence of a bullet hole in the side of the skull! This of course has changed everything, and he appears to have been relegated to the position of *anomaly!* An anomaly is something that cannot be explained by modern geology. Below is the picture of the skull with the bullet hole and artist's initial renditions. Now, in all fairness, the hole had been recognized by some but was relegated to *abscess damage*. The amazing thing about this "man," however, was that he appears to have been assigned a total of six different names by different writers. It just shows the subjective nature of interpreting evidence.

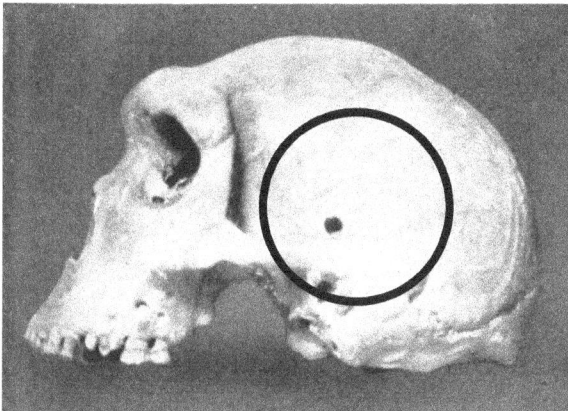

How many people had their faith in the Bible destroyed by this so-called scientific evidence for man's evolutionary history? Furthermore, artists' renditions of Rhodesia man only served to confuse the picture.

Nebraska Man

Shortly before the heated debate that took place in a small town in Tennessee in the 1925, now known as the Scopes Monkey Trial, a single molar tooth was found in 1922 in Pliocene deposits in Nebraska. Professor Henry Fairfield Osborn, head of the American Museum of Natural History, described the tooth as belonging to an early type of Pithecanthropoid (*belonging to the upright ape-man type*), which he named Hesperopithecus harold cooki after the geologist who discovered it. Hesperopithecus means, *ape of the Western world*. The discovery was published in the prestigious *Proceedings of the National Academy of Sciences*, as "Hesperopithecus, the First Anthropoid Primate Found in America."[36]

---

[36] "Hesperopithecus, the First Anthropoid Primate Found in America." *Proceedings of the National Academy of Sciences, USA* 8.8 (1922): 245-46. Print.

At the same time Grafton Elliot Smith, who had been involved with the Piltdown Man affair a few years before, persuaded the *Illustrated London News* to publish an artist's rendition of Hesperopithecus and his mate struggling together to survive the harshness of life. Shortly after the Scopes trial in 1928 it was discovered that a mistake had been made concerning the identity of the tooth of Hesperopithecus. The tooth was not that of man's evolutionary ancestor, but that of an extinct pig! Below are the pictures of the famous tooth and the artist's rendition of man's evolutionary link known as Nebraska Man.

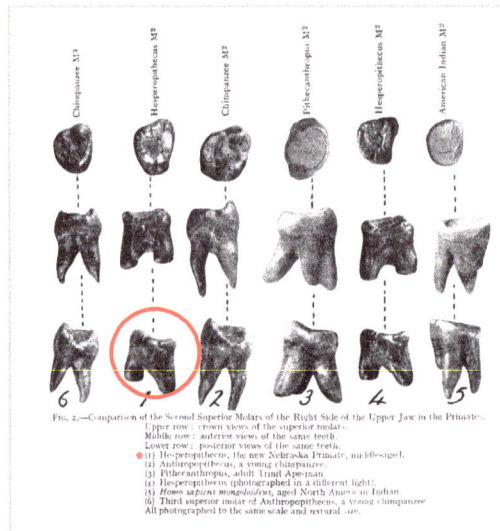

Nebraska man and wife all built up from one tooth? How many people might have had their faith in the Bible shaken because of this mistake?

Peking Man

During the 1920s, remains of what were claimed to be man's early ancestor were purchased in a Peking drugstore. Immediately these remains were dubbed, *Peking Man*, and called a missing link. Further inquiry produced the probable location of the remains – called *dragon-bone hill*. This place has been famous for all kinds of fossil discoveries including a mixture of the bones of man and other assorted ape-like fossils. But of course, these details were left out when introducing Peking Man to the public. To make a long story short, the remains of Peking Man were subsequently lost and all that remains of Peking Man is a plaster cast which adorns textbooks and Museums alike as more evidence of man's evolutionary history. Following is one of the few claimed photographs of the remains, the plaster cast and artists' renditions of Peking Man.

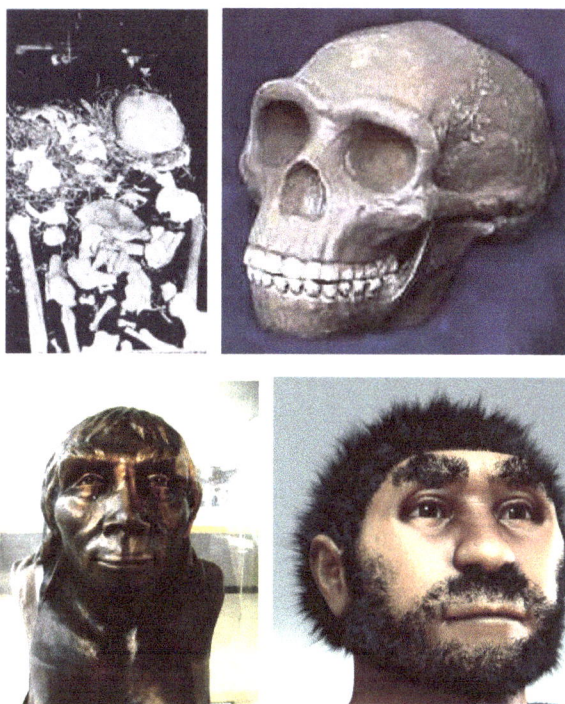

Peking Man – The remains have been lost but the biases remain! Artists' renditions helped to cement evolutionary dogma into the minds of men. How many people might have had their faith in the Bible shaken because of the presentation of Peking Man?

These are just a few of the hoaxes, frauds, and mistakes in the "science" of the evolution of man. I stop here because the late 1800s and the early 1900s were the formative years for evolution as a publicly accepted science. Everything we study today about fossil man or ape is predicated on those early discoveries. Those early discoveries continue to be seen by the public as evidence that the Bible is in error and that science has proven it. Most people do not know about those early mistakes, let alone the corrections. Producing this evidence and propagating it among the public has destroyed faith in the Biblical view of God and man - all due to preconceived ideas and manipulations of erroneous discoveries!

Since 1912 many discoveries of human-like and ape-like fossils have been discovered. Their remains have been carefully catalogued and all appear to be genuine. But here is the point – the fossils either show similarity to humans or similarity to apes. Even after 250 years of propagation of evolution and the search for fossils to prove it, no undisputed missing link has ever been found as proof for the evolution of man or any other creature. All of these fossils can be classified as human-like or ape-like. In other words, they can be categorized as fossils of man and fossils of extinct ape-like creatures, nothing more. Fossils are very rare and are open to interpretation – a lot of interpretation, most of it subjective.

Are these fossils evolutionary links or simply fossils from the Flood and post-Flood periods? Because of his rejection of history as presented in the Scriptures, man has blinded himself to the truth. Many have convinced themselves that they are descended form ape-like creatures and they continue to search for fossils to prove it. So long as this kind of biased search for missing links drives research, blindness will continue to cloud the judgment about human origins.

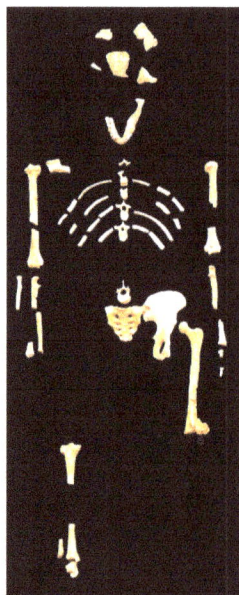

Lucy – everyone recognizes her. She has been on worldwide tours and has long been an icon for human evolution. (Many museums actually have a replica of these fossils.) Only three feet six inches tall, she resembled a *chimp*. She had a *chimp*-size brain, *chimp*-like teeth and *chimp*-like jaw. She had long, curved fingers and toes. Come to think of it, she was rather *chimp*-like. The only feature that pointed toward man was her knee, which was not found with the rest of the bones, and her pelvis. Could Lucy have simply been just a *chimp*?

The Neolithic Period

Archaeologists have come up with descriptive words like *Neolithic* that have been coordinated with The Geologic Time Table and the evolutionary view of man. These terms seem to communicate that the history of man has been identified and recorded and that these descriptive terms do indeed correctly identify the evolutionary progress of man.

The Neolithic Era, or Period (from the Greek language meaning *new stone*, or New Stone age, according to modern archaeologists), was a period in the development of human technology, beginning about 10,000 BC in some parts of the Middle East, and later in other parts of the world, and ending between 4,500 and 2,000 BC. This, of course, is the reckoning of modern dating methods which make several assumptions about the past by way of an evolutionary worldview. If the above description is true, however, then the Bible simply cannot be true. So we must take a look at this.

What I want you to see from this is not the dates, but the shift that took place with man during this unique period of history. If we ignore the dates for a minute, which are built on assumptions, then the picture that surfaces is one of a beginning of some kind – a new beginning. But where would we place this period of time in light of the Biblical chronology for man?

Archaeologists further describe the Neolithic as a period that commenced with the beginning of farming, which produced the *Neolithic Revolution*. It ended when metal tools became widespread (in the Copper Age or Bronze Age; or, in some geographical regions, in the Iron Age). The Neolithic is a progression of behavioral and cultural characteristics and changes, including the use of wild and domestic crops and of domesticated animals. The beginning of the Neolithic culture, according to archaeologists, is considered to be in what is called the Levant. It was the area that is identified as Jericho, the modern-day West Bank, about 10,200–8,800 BC. But how could this be if the Flood occurred 4,500 years ago?

Again, ignoring the dates, if we pick out the descriptive words, we come up with,
- New *(neo)*
- Stone
- Development of human technology
- West Bank
- Before the copper, bronze and iron age
- Beginning of farming

- Neolithic revolution
- Progression of behavioral and cultural changes

All these words describe a huge shift in culture and technology. This really could be describing the time immediately after the Tower of Babel event and for the next few hundred years. As people began to spread out from Babel, some would have taken technology with them. These would have gone on to build the Assyrian and Egyptian civilization. The Bible records that the Egyptians came from Noah's grandson, Mizraim. The Assyrians apparently came from Seth, Noah's son.

Some people would have been deficient in technology and perhaps took talents in art and music with them. These could have been people like the Neanderthals and Cro-Magnon people. These people would have struggled to exist, living in what shelter they could find, and expressing their life experiences in terms of art work on cave walls and in the remnants of musical instruments.

It would have been a new beginning for man. Without the cooperative efforts of what had existed at Babel, technological skills lost would have needed to be learned again. Some would have done fine with this, others would not have fared so well. Some of the Neolithic tools found reflect this condition. Some are very sophisticated, while others appear to be crude. The so-called Neolithic Period could really be the period of time immediately following the spreading out from Babel. If we adjust our thinking to be in line with the Biblical chronology, then the discoveries of artifacts like Neolithic tools could easily be explained. It is a different way of interpreting the evidence, dates and ages aside.

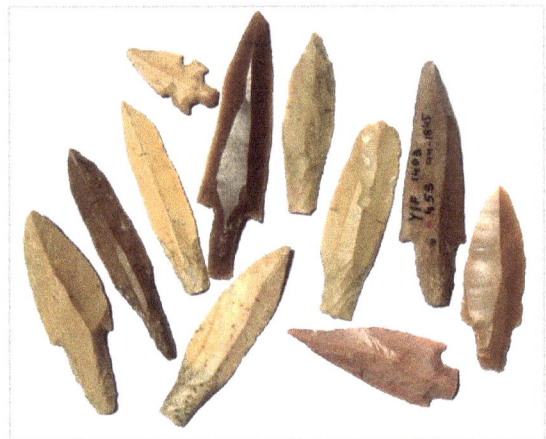

Stone tools of the Neolithic Period – some showing a high degree of technological know-how.

Neolithic cave art   Neolithic farming equipment   Neolithic dishes and cookware

All these implements could have been developed and used by struggling people to survive after the Tower of Babel.

Activity 10

Choose at least four more fossil "man" discoveries since the early 20th century that have been used to support the evolution of man. Write a report on these finds including discoverer, place of discovery, date of discovery, initial announcement, and any corrections or realignments since their discovery.

Please take Quiz #10, Appendix B

Lesson 11 – The Biblical History of Man

A Short Overview

Words to know: *proclivities   chronology   deluge   hieroglyphs*

The true history of man from Creation to the establishment of Israel as God's chosen nation is presented in the Book of Genesis – the book of beginnings. Without this history man has nowhere to turn for genuine answers to his unique existence. Without the Genesis creation, man has no identity. Without the Fall, man has no explanation for his **proclivities**, sin, and need for a savior. Without the Flood, man has no explanation for the geology of the earth or for the act of judgment by a righteous God who holds man accountable for his choices. Without the Tower of Babel incident, man has no explanation for the appearance of languages. And without Abraham, man has no answer for the uniqueness of Israel as a people that have survived for thousands of years despite repeated attempts to wipe them out.

The Biblical story of the Tower of Babel has been relegated to the realm of myth. But without it, we have no explanation for different languages or an explanation for the existence of cave men and the spreading out of civilization. We are left with an evolutionary, naturalistic explanation of man which certainly is no explanation for all that we have come to cherish about man's uniqueness and moral compass.

Secular dating is all off

One of the areas of controversy for the Bible believer involves the conflict between Biblical **chronology** and secular chronology. The Biblical chronology of the history of man is a fairly easy one to figure out by using the genealogies in Genesis chapter 5. But it has been rejected by modern archaeology. And because it has been rejected, an alternative chronology had to be found. Evolutionary views had already placed the advent of man at many thousands of years earlier than the Bible's chronology. So of course, the Biblical history could not be right. The chronology adopted by secular historians has been the Egyptian chronology. There is a fascination with ancient Egypt among archaeologists. And there is a prevailing view that the Egyptian chronology is the "gold standard" of chronologies. Anything that differs from it is suspect. So, what does all that have to do with the Bible and the history of man?

Secular historians tell us that the First Dynasty of Egypt began in BC 3150 and its pre-dynastic period goes back to BC 5000. But using the Biblical chronology places the Creation at BC 4000, the Flood at BC 2344, the Tower of Babel incident at BC 2244, and the earliest Egyptian civilization at some time after BC 2244. There is a huge discrepancy in these dates! Either there is an error in the Egyptian chronology or there is an error in the Bible. But if there is an error in the Bible, then none of its claims on mankind are true and its teachings about God, a coming Savior and eternal punishment are false!

Secular historians are so convinced of these dates that statements like the following are common place when it comes to the Bible.

> David and Solomon are portrayed in the Bible as two of the greatest kings of the ancient world, yet within the conventional (the secular) chronology, a suitable context for their reigns cannot be found. The Bible is the only written source concerning the United Monarchy, and it is therefore the basis of any historical presentation of the period.[37]

Up until the late 1700s Western Civilization respected the Biblical history of man including the six-day creation and global flood. Take a look at the chart and the plate below. They may surprise you! They are from the Encyclopedia Britannica of 1771! Even the famous Encyclopedia Britannica embraced a young Earth and global Flood at one time!

---

[37] Mazar, Amihai. *Archaeology of the Land of the Bible.* Anchor Yale Bible Reference Library, 1990. Print.

A Table of Remarkable Eras and Events	Year of the World	Before Christ
The Creation of the World	0	4007
The Deluge or Noah's Flood	1656	2351
The Assyrian Monarchy Founded by Nimrod	1831	2176
The Birth of Abraham	2008	1999
The Destruction of Sodom and Gomorrah	2110	1897

The chart above is a recreation of the table that appears in the Encyclopedia Britannica, 1771, page 493, under the heading, *Astronomy*. The article treats these events as actual historical events and times.

Encyclopedia Britannica, 1771, page 414, under the heading, "Deluge", features the above plate depicting "Noah's Ark, floating on the waters of the Deluge." The Encyclopedia says of this event,

"... the most memorable (event) was that called the universal deluge, or Noah's flood, which overflowed and destroyed the whole earth, and out of which only Noah, and those with him in the ark, escaped."

What brought about the radical departure from this view? The Enlightenment! The Enlightenment was primarily connected to the 1700s and 1800s and was the key event that produced a shift in the way man thought about himself, God and the world around him. From lesson 8 of our study, we discovered that The Enlightenment produced the following changes:

- The belief that man, apart from any biblical revelation, could discover his past
- The belief that the Bible is at worst a myth and at best a corrupted record of the past
- Gave us the religion of Deism; God created but is not involved – God is irrelevant
- Gave us the idea that scientific discovery alone could uncover our past, including Egyptian History (the greatest and oldest civilization of man), not reliance on the biblical revelation; (By scientific discovery, they include their interpretation of the artifacts as determined by their naturalistic bias – the Bible cannot be considered)

The breakthrough concerning Egyptian history came in 1799 with the discovery of the Rosetta Stone. Up until that time Egyptian *hieroglyphs* were a mystery. No one could figure out what they meant. The discovery of the Rosetta Stone changed all that. It consisted of one inscription written in three languages – Egyptian hieroglyphs, ancient Egyptian and Greek. The man of the hour was Jean-François Champollion (1790 –1832), who became known as *The Father of Egyptology*. He provided the key to understanding the ancient hieroglyphs because he knew both Greek and ancient Egyptian. All of a sudden, the door was opened to Egyptian archaeology like never before. Hundreds of thousands of ancient artifacts were smuggled out of Egypt in an effort to understand ancient Egypt and its mysteries.

Now considered to be the Father of Egyptology and a respected authority on translating the ancient documents, he went to work translating other inscriptions. One in particular was an Egyptian hieroglyph mentioning the Egyptian pharaoh Sheshonk I. Champollion, whether by preconceived ideas or through a translation mistake, misidentified this pharaoh. He identified him as the Biblical pharaoh, Shishak in 2 Chronicles 12. This one seemingly small mistake threw the whole of Egyptology off by as much as 600-1,000 years. Of course, since Egyptian Chronology had become the "gold standard" of ancient history, no discoveries of Biblical personages, inscriptions or artifacts made any sense in light of Egyptian history which had been built off of Champollion's mistake. This had the effect of confirming the belief that the Bible was just a bunch of myths. The translation error was finally caught in 1888, but the misidentification remained intact.

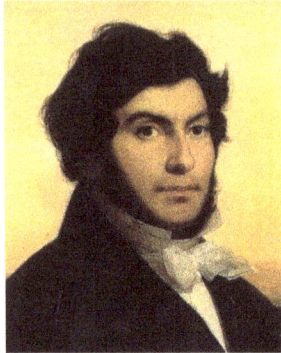

Champollion and the Rosetta Stone

Ever since then the Bible has been viewed as an ancient mythical story that has been in error in its calculations of man's past. None of the Biblical artifacts discovered seemed to fit with any of the ancient chronologies.

Many archaeologists today have started to correct this mistaken idea and have attempted to adjust for the errors by readjusting the misidentification of Sheshonk I with Sheshak who lived during King Rehoboam's time. Needless to say, this attempt has come under a lot of fire. Firmly held beliefs die hard.

Some archaeologists have become aware of the fact that because of some of these corrections, many of the Biblical artifacts discovered earlier which had been relegated to a different time, suddenly fit into history! Of course, these archaeologists are in the minority today. The Bible is still ridiculed among the majority of archaeologists as just another myth that was copied from other sources and cultures.

Think about this – if the Bible was accepted as historical and reliable, think of the amount of change that would have to take place in not only archaeology, but in anthropology, biology, and in geology! The shift back to the Bible would be catastrophic for the modern sciences and for the gains of The Enlightenment! The Enlightenment was supposed to have delivered man from the bigotry and myths of the Bible!

Archaeologist David Rohl can account for as much as 1200 years of errors in the Egyptian Chronology! If these errors are corrected, the history of Egypt falls more in line with Biblical history! The many Biblical artifacts discovered over the past 150 years now make sense and verify the Bible as reliable. Here are just a few of the discoveries.

- Evidence of Abraham in Egypt
- Evidence of Joseph in Egypt
- Evidence of the slaughter of the Israelite males
- Evidence Joseph as second in command of Egypt during a famine
- Evidence of Israel's rapid departure from Egypt
- Evidence of the plagues of Exodus
- Evidence of Moses' high position as "son" of Pharaoh
- Evidence of David
- Evidence of Solomon and his Egyptian wife

In fact, in his book, *Pharaohs and Kings, A Biblical Quest*, David Rohl lists 40 archaeological discoveries that would fit history perfectly if the miscalculations in the Egyptian Chronologies were corrected. What is interesting about this is that archaeologist David Rohl is agnostic in his belief but is convinced of the Bible's historical accuracy. If you can obtain a copy of his book, it is well worth the expenditure.

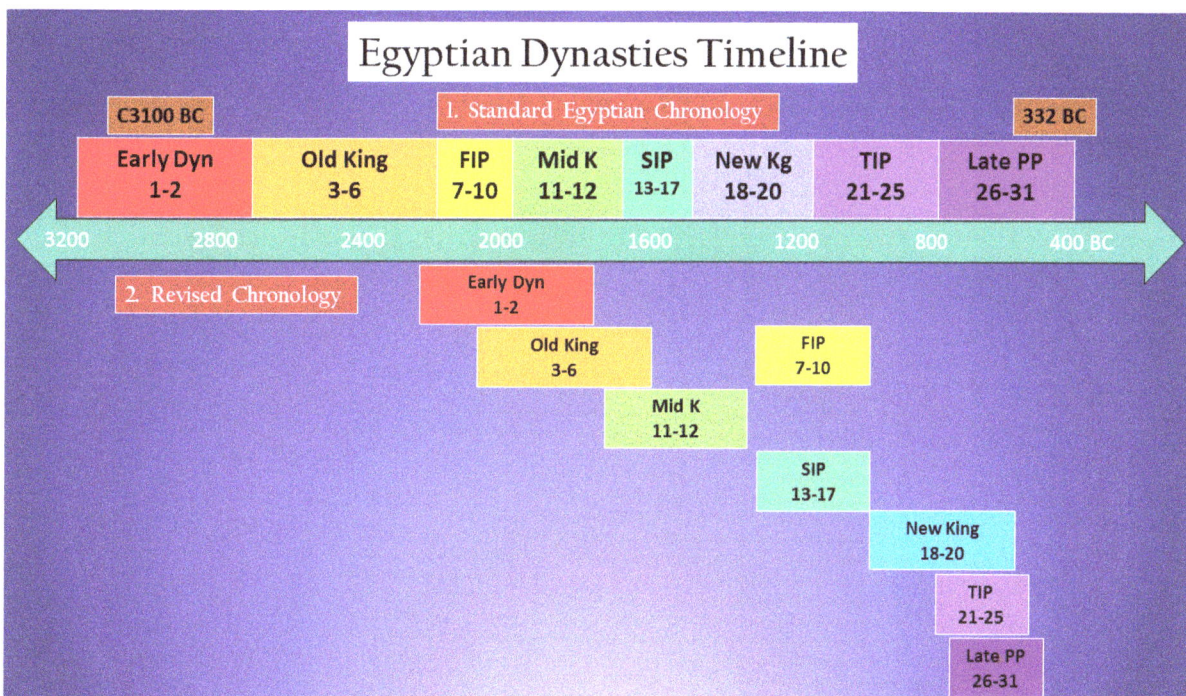

David Rohl's adjusted Chronology – The abbreviations stand for the different periods of Egyptian history that have been constructed by archaeologists. You can see that if the adjustments are taken into account, the different periods of Egyptian history change by as much as 1000 years. Now the Bible and Egyptian history seem to fit well together and there are no contradictions.

163

BANQUE MISR

بَنْك مِصْر

نعمل معاً لخير بلدنا

King Menes (in the secular language) is Mizraim (in the Hebrew). Mizraim translated into the Greek is Egypt. He was the grandson of Noah, the first king of Egypt who would have established the 1ˢᵗ Dynasty of Egypt shortly after the Tower of Babel. If the miscalculations are corrected, then he fits just where the Bible would place him! The image above right is of the logo of the National Bank of Egypt – Misr, after Mizraim, Noah's grandson!

In 2001 fourteen wooden boats were discovered in Egypt from the 1st Dynasty. The boats were around 81 feet long. Naval archaeologist Cheryl Ward said she was amazed by the high degree of technical skill shown by these artifacts. They were sealed in mud brick casing and buried intact. This is significant because later boats buried with the pharaohs were dismantled. Being so close to the Flood in time, could these boats have been buried with the thought that they might just be needed in the afterlife for another flood? Look for the article on this fascinating discovery at https://www.abc.se/~pa/mar/abydos.htm.

Secular archaeologists tell us that Djet (Zet) was the fourth pharaoh (king) of the 1ˢᵗ Dynasty. Above is the stele of King Djet. He took as the sign of his name, the serpent. Could this correspond to the serpent of the Garden? Is this, perhaps, a reminder of the story of the garden, in honor of the one who may have helped him to gain power?

In 1922-1934 Sir Leonard Woolley (right, with T.E. Lawrence) discovered Ur as the first civilization with a superior knowledge of astronomy and arithmetic. If the miscalculations are corrected in the chronology, Terah, the father of Abraham, would have been living at the time of Zoser (Djoser) of the 3rd Dynasty of Egypt. The step pyramid of Zoser shows the influence from Ur in Mesoptamia where many ancient step pyramids have been discovered.

If miscalculations are corrected in the chronology, then the whole idea of pyramids connected to the Tower of Babel fits very well.

This is King (Pharaoh) Khufu (Cheops in Greek) and his pyramid. If the miscalculations are corrected in the chronology, then Khufu would fit during the time of Abraham's sojourn in Egypt. I don't think it is any coincidence that it was during Khufu's reign that the true pyramid was built. Abraham could very well have brought a more developed mathematics with him to Egypt from Ur, where he had originated. This would have aided the construction of a true pyramid during Khufu's reign.

The pyramids before Khufu were inferior and not square.

If the miscalculations are corrected in the chronology, Joseph would most likely have lived during the reign of Sesostris I who was known to have had a vizier (prime minister) named Mentuhotep. This vizier had exceptional ruling power. "…Mentuhotep…appears as the alter ego of the king. When he arrived, the great personages bowed down before him at the outer door of the royal palace." [38]

[38] Bey, Heinrich Brugsch. *Egypt Under the Pharaohs*. Whitefish, Montana: Kessinger, 2004. Print.

(Left) This statue was found in 1987 in Egypt. The personage is not Egyptian. He has reddish-blond hair, and is dressed in a style not typical of an Egyptian. He is wearing what appears to be a *coat of many colors*, which also was not typical of Egyptians – could this be a statue made in the likeness of Joseph (Mentuhotep)?

(Right) The Nile River, with a portion of Joseph's Canal visible running parallel to the Nile, and extending into the Faiyum Oasis: this canal, still in use today in Egypt, is called in Egyptian, *Joseph's canal*. Could this be the remnant of what Joseph had built for Egypt during the years of famine recorded in Genesis?

There are many of these kinds of things that do not fit into the secular chronology because it is off in its dates due to apparent mistakes made through the years. If the miscalculations are corrected and the pharaohs' reigns are adjusted accordingly, then many of the archaeological artifacts of the Bible that have been put on the shelf because they were thought to have been found in the wrong time, now suddenly come to life.

The Bible gives man all he needs to know about life, his origin and his destiny. Below is a brief outline of God's history of man. See how much you already know as you work through this outline.

The Bible's history

A brief outline of history can be structured using the main points from each section of Scripture:

1. The Creation

a) The creation of the space and Earth from nothing. This would include the elements. Most of the elements on the Periodic Table of Elements have been discovered in every part of the universe that has been explored with telescopes and other sophisticated equipment. This shows design and an all-powerful Designer

b) The creation of plants

c) The creation of the sun, moon and stars – after the space and Earth. This runs totally contrary to the Big Bang idea which says that the stars came first, then the sun, the Earth and finally the moon out of the Earth

d) The creation of sea creatures and birds

e) The creation of land creatures and man and woman

2. The Fall

a) The temptation by the serpent who had infiltrated God's garden.

b) Disobedience by Adam and Eve

c) The beginning of the corruption of the Earth and Adam's offspring

3. The Flood

a) The ruin of the pre-flood Earth by man's sin

b) The global flood which killed all land-dwelling animals and man which were not on board the ark

c) The salvation of Noah, his wife, his three sons and their wives, along with selected animals from the pre-flood world

d) The adjustment of the post-flood Earth bringing on an ice age lasting from between 500 and 700 years

4. The Tower of Babel Incident

a) Man's disobedience again brought God's judgment

b) The confusion of speech into language groups and the consequent dispersion of man

c) The beginning of the civilizations of Egypt and Assyria; the isolation of certain tribes of people such as Neanderthals, Cro-Magnons and other cave dwellers

5. The selection of Abraham as the father of a nation that would bring about blessing upon all mankind via a savior

a) The calling of Abraham and Sarah to go to a place which God would give to Abraham's descendants

b) The promise of a son – Isaac – through whom the blessing would com

c) The miraculous birth of Isaac in fulfillment of God's promise

d) The birth of Esau and Jacob and the passing of the promise to Jacob and his descendants

e) The birth of Jacob's sons, one of which would bring salvation to Jacob's tribe now known as Israel

6. The deliverance of Israel from Egypt

a) Moses' leadership

b) The establishment of the Law and a special tribe that would produce priests who would make atonement for Israel while God was in their midst and until the real High Priest would come

c) Joshua's leadership; the giving of the land promised to Abraham

d) The apostasy of Israel and division into two countries, the Northern Kingdom known as Israel and the Southern Kingdom known as Judah

e) The judgment of Israel by God upon the Northern Kingdom through Assyria

f) The judgment of Israel by God upon the Southern Kingdom through Babylon

g) The return of Israel to the land that had been promised by God

h) The coming of the long-awaited Messiah – Jesus who would save His people from their sins

i) The rejection by Israel of God's appointed Messiah

k) The offering of salvation through the Jewish Messiah to the Gentiles

l) The judgment of God upon Israel through the Romans

7. Now what? The Scriptures are full of prophecy concerning both the church and the nation of Israel and the fulfillment of the rest of God's promises regarding man.

a) The restoration of Israel to its promised homeland

b) The coming of the anti-Christ who will attempt to thwart God's plans for Israel

c) The second coming of Christ to save His people, the Jews and to establish His kingdom on earth

d) The 2nd attempt by the Serpent to finally and totally defeat God's plans for man

e) Satan is finally destroyed; the final judgment; and the total destruction of the present heavens and earth and creation of a new heavens and earth

f) The ushering in of eternity and eternal paradise for the true believers

God's kindness and care for man has been demonstrated time and time again throughout history. The fact that He has revealed the general plan for man is one of the kindest things He could have done for man who has demonstrated time and again throughout his wretched history that he wants nothing to do with the God of the Bible. If we do away with the Bible's history, then man truly has no historical perspective of where he came from or where he is going. It is a miserable picture of despair and futility as man continues to suppress God's truth for his own wisdom gained in The Enlightenment.

Activity 11a

Using Genesis chapters 5 and 11, see if you can construct a chronology of Biblical history from Adam to Abraham. How would you begin? What guidelines will you use in calculating the number of years involved from Adam to Abraham? On your chart be sure to include the major events included in that period of time.

Activity 11b

In you kit, locate the small bag of stone tools, labeled *Neolithic stone tools*. Read the enclosed description, and describe why it might have been necessary for people to learn how to use and craft stone tools. Why might some people have done well, and some people have struggled after the dispersion from the Tower of Babel?

Please take Quiz #11, Appendix B

Appendices

Appendix A – Fossil Anomalies

Whenever a discovery is made that does not fit the standard thinking of modern geology, it is labeled as fraud, grossly in error or as an anomaly. An anomaly, in geology, is something that cannot be explained by modern principles as geology currently defines them. The list of these is in the hundreds. The late William Corliss spent a lifetime accumulating scientific articles on these anomalies that had been published by accepted scientific journals and magazines. They are fascinating and they show us that things that cannot be explained by modern science are easily explained by Biblical history. Below are a few of these anomalies.

Insects have not changed over hundreds of millions of years. Other than giant insects appearing in Pennsylvanian time between 320 and 290 million years ago, they look the same.

Geologic timetable; Dragonfly with wing span up to 28 inches

Discover magazine, **January 1992**, pages 30-31, reported living bacteria in a Mastodon's stomach found in Ohio. The Mastodon is supposed to be 11,000 years old. How could bacteria live that long? Biologists insist that bacteria could not have survived for that length of time.

Fossil Tree stumps (Sequoias) 10-12 feet across in Florissant, Colorado have been discovered that show evidence of having been sheared off. Scientists tell us that this happened in a mud flow. It would have to have been a flood of high velocity and force to accomplish this. The mud flows were accompanied by ash deposits which preserved countless insects along with the stumps.

Coelophysis, a Triassic theropod from the Ghost Ranch area in Texas – over a thousand skeletons have been found from this area alone. This is considered to be an anomaly because it does not depict natural death. This was undoubtedly a catastrophic burial.

Two coelophysis specimens buried together

Sturgeon – evidently have been around for 250 million years, yet unchanged!

Tuatara – living fossils that have been around for 200 million years, unchanged!

Dinosaurs and birds together in the same fossil strata! Dr. Paul Serrano from the University of Chicago said this on the topic of evolution: "What is becoming apparent is that many of the modern bird groups – parrots, maybe even penguins, and other kinds of groups like owls – evolved earlier in the dinosaur era, and we are beginning to pick up their traces." [39]

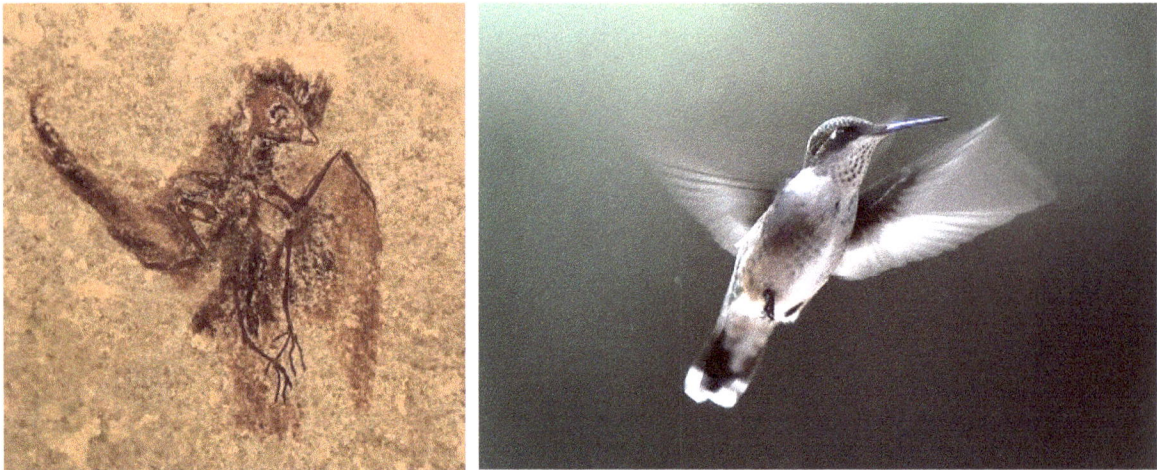

Evolutionists believe that this fossil bird is some way related to hummingbirds and swifts. It could have fit in the palm of your hand.

Long-Buried, Undecomposed Organic Matter – There are numerous records of animal and plant material, in relatively fresh condition, found buried in glacial debris, muck, and soil, dated from 10,000 to millions of years old. Some of the animal material seems quite fresh. Vegetation is often green, and the wood burnable. Some of the skeletal bones are associated with skin, tissue, marrow, etc. The evidence would lead one to believe that the modern geological dates given for these remains are simply wrong. These remains cannot be that old, as things such as skin and tissue quickly decay. Many of these examples have been reported in William Corliss' books on anomalies in geology. Here are just a few selections from the science articles that he collected:

- The whole of northeast Siberia is one vast graveyard filled with the bones of animals that perished within comparatively recent times…The whole of northern Siberia, from the Ural Mountains to Bering Strait, is one vast graveyard filled with animal remains…These bones occur everywhere…the lower portions of the icy chasms are filled with tusks, bones and skulls in countless abundance…Still more amazing is the fact that the islands in the Arctic Ocean north of Siberia are equally full of the tusks and bones of elephants and rhinoceroses…Stranger still, actually the very bodies of these great elephants, with flesh, fur and hair perfect, are seen standing upright in the frozen cliffs…after being entombed for thousands of years, that the wolves eat the flesh. "The Glaciated Grave of the Mammoth in Siberia," *Current Opinion*, 61:330, 1916.

---

[39] Werner, Paul. *Living Fossils, Evolution: The Grand Experiment*. Vol. 2. New Leaf, 2008. 164. Print.

- Even conventional geologists have to marvel a bit at the presence of frozen mammoths and rhinoceroses, thousands of years old, in the Arctic muck. "The flesh is as fresh as if recently taken out of an Esquimaux cache or a Yakut subterranean meat-safe." Howorth, Henry H. "The Sudden Extinction of the Mammoth," *Geological Magazine*, 2:9

- :309 and 2:8:569, 1881.

- Dr. Herz then removed the skull, and found the well-preserved tongue hanging out of the mandible. He also noticed that the mouth was filled with grass, which had been cropped, but not chewed and swallowed. Further examination of the carcass showed that the cavity of the chest was filled with clotted blood. "The New Mammoth at St. Petersburg," *Nature*, 68:297, 1903.

- Here, on the contrary, we have remains of whole herds together; the bones equally preserved, the ivory equally fresh, and pointing to but one conclusion, that they perished in herds where they are found, and perished by some overwhelming cataclysm. The fact of so many of the remains being found in high ground seems to show that this high ground was a place of refuge where the beasts congregated in the presence of some common danger, such as a general inundation which threatened to annihilate them. In this way also we can best account for the heterogeneous character of the collections of bones, mammoth and rhinoceros, bison...musk sheep and stag, etc., animals that do not naturally herd together...the presence of bones of other large mammals over 100 miles from the mainland: all over the hills in the interior of the island...occasionally the bones of other Pleistocene animals: Arctic hare, wolf and wolverine (including one whole wolverine skeleton with skin and hair still attached to its head and paws), horse, bison, reindeer, wholly rhinoceros and cave lion. Stewart, John Massey. "Frozen Mammoths from Siberia Bring the Ice Ages to Vivid Life," *Smithsonian Magazine*. 8:61, December 1977.

- The presence of buried trees and logs, still fresh enough to burn, have been found in association with the mammoth remains. Howorth, Henry H. *The Mammoth and the Flood*, London, 1887

- **Alaska** – ...the Alaskan muck is heavy with organic debris, much of it very fresh-appearing...Mammal remains are for the most part dismembered and disarticulated, even though some fragments yet retain, in this frozen state, portions of ligaments, skin, hair, and flesh...Practically all the organic material has come from grasses, mosses, alder, spruce, willow, cottonwood, and birch...Much of the vegetable material is in place, with stumps of trees still embedded and upright; in one 20-foot section, six horizons of residual vegetation were plainly evident...Vertebrate remains – usually bones, but sometimes almost complete skeletons, and occasionally bones or skeletons with skin and flesh adhering – occur throughout the muck. These remains include living species such as moose, caribou, and many smaller types, and extinct species such as mammoth, mastodon, saber-toothed tiger, super-bison, and camel. "Much about Muck," *Pursuit*, 2:68, October 1969.

- **The Canadian Arctic** – Axel Heiberg Island is on the Arctic Ocean less than 700 miles from the North Pole. Dense forests seem unlikely at such high latitudes, but exploration of the island by J. Basinger, of the University of Saskatchewan, has revealed the stumps of a 45 MYO forest...found himself in a time-frozen, once-lush forest similar to the present Cypress Swamp in Florida's Everglades. He estimates that some of the trees could have been as tall as 150 feet. Some of them remain rooted in the ancient soil amid a debris of leaves above the rock-hard permafrost. "We packed the leaves into a bag. They're like a handful of fresh leaves except they're blackish, a bit brittle. The leaves and the wood are incredibly fresh, even though dated geologically at 45 MYO...The forest was indeed dense: the stumps are only about ten paces apart, and some are as much as six feet across...they must have been killed off relatively quickly for the roots not to decay...the stumps and logs of Axel Heiber Island are not fossilized, rather they cut like fresh lumber and can be burned." Howse, John; "Forestry Frozen in Time," *Maclean's Magazine*, p. 55, September 8, 1986.

- **Central United States** – In this region, deeply buried, undecayed wood is rather common. Mammoth bones are also found, occasionally with remnants of skin and sinews attached. **Indiana** – A Mastodon found near Covington, Indiana...The teeth were in a good state of preservation, and Mr. Perrin Kent states that when the larger bones were cut open, the marrow, still preserved, was utilized by the bog cutters to grease their boots...Near Iroquois County, Indiana, on inspecting the remains (of a Mastodon), a mass of fibrous, bark-like material was found between the ribs, filling the place of the animal's stomach; when carefully separated, it proved to be a crushed mass of herbs and grasses, similar to those which still grow in the vicinity. "On the existence of the Mammoth in Recent Times in North America," Geological Magazine, 2:8:373, 1881.

- **Indiana** – N.H. Winchell, in 1875, issued a long report on the vegetable remains buried in the glacial drift...In Franklin County, Dr. Rufus Hammond mentions, that in digging wells on the uplands, the roots and bodies of trees are frequently found at various depths, from ten to thirty feet; and occasionally limbs and leaves are found with vegetable mould, at various depths. Corliss, William, ed. *Geological Anomalies.*

- **Cincinnati, Ohio** –... a well near the corner of Fourth and Vine Streets, Judge Burnet, who was then the proprietor, struck at the depth of ninety-three feet, a partially decayed stump, with the roots attached, standing in an upright position. Whittlesy, Charles; "Notes upon the Drift and Alluvium of Ohio and the West," American Journal of Science, 2:5:205

- **New Jersey** – An industry the like of which does not exist anywhere else in the world furnishes scores of people in Cape May County, New Jersey, with remunerative employment, and has made comfortable fortunes for many citizens. It is the novel business of mining cedar trees – digging from far beneath the surface immense logs of sound and aromatic cedar...These ancient trees are of a white variety of cedar, and when cut have the same aromatic flavor intensified many degrees that the common red cedar of the present day has...This has been going on for 3/4 of a century. "Subterranean Woods," Scientific American, 52:22, 1885.

- **California** – A gold miner wrote to *Scientific American* in 1871 about unpetrified wood found at great depths in California gravels...They have been found in great numbers and quite large...More amazing was a buried log found near Portola, California. A redwood log discovered near the Oroville Dam has been determined to be 10 MYO and still capable of burning. "Fossil Trees in California," Scientific American, 25:5, 1871; "Log Rated 10 Million Years Old," Phoenix Gazette, May 6, 1974.

- **England** – At a depth of 40 feet below the level of the adjoining land, trees (chiefly oak) are found in all positions...One oak tree...in the thickest part measuring 12 1/2 feet in circumference...the trees cannot be less than 3,000 years old; and would require at least 300 years to attain the dimensions given. "On the discovery of Ancient Trees below the Surface of the Land at the Western Dock Now under Construction at Hull," Report of the British Association, 1866, p. 52.

- **Great Lakes Region** – In the August 17, 1883, issue of *Science*, J.W. Dawson listed some "unsolved problems of geology." Among these puzzles were the buried leaf beds on the Ottawa River and the drift trunks found in the boulder clay of Manitoba. <small>Dawson, J.W. *Science*. August 17, 1883.</small>

- **Michigan** – In 1976-77, spruce and tamarack trees in growth position were exposed between 18 and 33 feet below the surface…The wood of these trees was remarkably well preserved, with only the bark and less than an inch of the outer layer showing any carbonization. Similar buried forests have been found at Two Creeks, Wisconsin, and Cochrane, Ontario. <small>Buried Forest Tells Glacial Tale," *Science News*, 113:229, 1978.</small>

- **New York** – In 1872, near Otisville, New York, a large mastodon was unearthed. In its stomach was found a quantity of undigested matter. Among it were fresh-looking and very large leaves, of odd form, and blades of strange grass, of extreme length, varying from an inch to three inches in width, and looking as if freshly cropped from the earth. <small>"A Gigantic Mastodon," *Scientific American*, 26:264, 1872.</small>

- **South America** – "Abundance of Pleistocene animal fossils…compares very well with Siberia and North America…" Charles Darwin observed, "…some remains of a very large unknown mammal exhumed from Pleistocene deposits…appeared so fresh that…it is difficult to believe that they have lain buried for ages underground. The bone contains so much animal matter, that when heated in the flame of spirit-lamp it not only exhales a very strong animal odour, but likewise burns with a slight flame." <small>Hapgood, Charles H. *The Path of the Pole*, Philadelphia, 1970, pp. 249 and 280.</small>

Appendix B - Quizzes

Quiz #1
1. The foundation of modern geology is called _____.
2. What book influenced Darwin in his writing of *Origin* in 1859?
3. The word fossil means _____.
4. Name two ways we can know about the past.
5. What is the difference between theism and deism?

Quiz #2
1. Briefly describe why all systems in a living thing must be present in order for the living thing to survive.
2. Briefly describe the Cambrian Explosion. Why is it significant?
3. List three living creatures that are typical of the Cambrian rock layers.
4. What is the Ediacaren? What is the significant about it?
5. Describe one main difference between the order of creation in Genesis and the order of the development of life in evolution.

Quiz #3
1. What was the cause of the Genesis Flood?
2. What makes the Bible a unique set of writings?
3. Name two ways that man attempts to discern his future?

Quiz #4
1. How do bone beds and fossil graveyards support a global flood?
2. How might the global flood of Genesis explain the general order of fossils in the rock layers?

Quiz #5
1. Briefly describe the Geologic Time Table.
2. Briefly describe how the Geologic Time Table was developed.
3. What is an *out of place fossil*?
4. What is an *index fossil*?

Quiz #6
1. Briefly describe the exceptions to the evolutionary ideal of correlating the rock layers.
2. Briefly describe what index fossils are, and what they are supposed to indicate.

Quiz #7

1. Who was the inventor of the word, *dinosauria*?

2. What were dinosaurs called before 1842?

3. Dinosaurs are divided by paleontologists into what two groups? What do these names mean?

4. Thomas Huxley thought that Archaeopteryx was a combination of what two animals?

Quiz #8

1. Darwin was famous for his study of _____.

2. What did Darwin's study show?

3. Pick one evolutionary transition and describe the kinds of changes that must take place for one kind of creature to evolve into a different kind of creature.

Quiz #9 - Briefly describe the Biblical view of man.

Quiz #10

1. Haeckel was famous for his _____ _____.

2. Pick one example of a fossil mistake in the evolution of man and describe the story behind it.

Quiz #11

1. Briefly outline the history of man from God's perspective.

2. What was the crucial mistake that Champollion made that affected Egyptian chronology?

3. If secular dating of the Egyptian dynasties is correct, what are the implications for the veracity of the Bible? Why?

Appendix C – Final Test

Final Test – Matching

Match List 1 with List 2 by putting the appropriate letter in the List 1 blank from List 2.

List 1
1.____ Deciphered the Rosetta Stone
2.____ The one who popularized uniformitarianism
3.____ Belief that God is not involved in the affairs of the earth
4.____ Richard Owen
5.____ Thomas Huxley
6.____ James Hutton
7.____ *lizard-hipped*
8.____ *bird-hipped*
9.____ Dubois
10.____ Pig's tooth
11.____ Hoax originated in England
12.____ Evolutionary arrangement of the rock layers
13.____ *dug up*
14.____ *missing links*
15.____ A tall man who lived in caves
16.____ Bullet hole!
17.____ Missing bones
18.____ Ex nihilo
19.____ Finch
20.____ More imagination than scientific fact!
21.____ Birds supposedly evolved from this
22.____ Land creatures (mammals, reptiles, dinosaurs, creeping things and man)
23.____ 600th year of Noah's life
24.____ Dispersion and confusion
25.____ Ice Age
26.____ Egyptian history
27.____ Bone beds and graveyard fossils
28.____ Cambrian Explosion
29.____ Ediacaren
30.____ Crinoid

List 2

a) Masses of fossil bones or shells buried together

b) First multicellular life

c) Bottom rock layers with bizarre looking fossils

d) Sea Lily

e) Champollion

f) Lyell

g) Deism

h) Inventor of dinosaurian

i) Beginning of the Genesis Flood

j) Tower of Babel incident

k) A post-flood event

l) Considered the "gold standard" of chronologies

m) Promoted Archaeopteryx as missing link

n) Earth history is cyclical

o) Saurischian

p) Ornithischian

q) What Darwin studied

r) Haeckel's drawings

s) Theropod

t) 6th Day of Creation

u) Java Man

v) Nebraska Man

w) Piltdown Man

x) Geologic Time Table

y) Fossil

z) Gaps between kinds

aa) Cro-Magnon Man

bb) Rhodesia Man

cc) Peking Man

dd) out of nothing

Appendix D – Answers to the Quizzes

Quiz #1

1. uniformitarianism
2. *Principles of Geology*, by Charles Lyell
3. *dug up*
4. Any two of these: archaeological artifacts, eye witness accounts, historical records, fossils
5. Theism involves a God who is involved in His creation through miracles. Deism involves a god who is absent and uninvolved in the creation.

Quiz #2

1. Systems are interdependent. They all work together, or they don't work at all. They do not function independently of one another.
2. The Cambrian Explosion is a description by paleontologists to describe the sudden explosion of life and its diversity in the bottom-most rock layers that have multicellular life. The ancestors to this life are not in the rock layers below.
3. Any three of these: trilobites, corals, bryozoans, brachiopods, mollusks, crinoids, sponges, worms, chordates.
4. The Ediacaren describes rock layers originally discovered in Australia supposed to be older than Cambrian rock layers and in which fossil oddities were found and claimed to be the supposed ancestors of the Cambrian fauna.
5. Any one of the following: space and earth before sun, moon and stars; plants before the sun; plants before sea creatures; birds before dinosaurs; mammals, dinosaurs, and man living at the same time; creation over 6 literal days, evolution of life over a couple of billion years.

Quiz #3

1. The cause of the Genesis flood was God's judgment on the wickedness of man. It was not a natural event, but one in which God was directly involved and initiated.
2. The Bible is more than the writings of man. It is a compilation of revelation of God letting man know where he came from, what happened to him and why he needs a savior.
3. Any two of the following: divination; the use of physical objects to aid in discerning the future or the past, witchcraft; the use of spells and incantations to aid in discerning the future or the past, omens or sorcery, casting spells, mediums or spiritists or necromancers (those who contact the dead to gain insight into life or the unknown)

Quiz #4

1. Bone beds and fossil graveyards are large deposits of disarticulated shells and bones that were buried catastrophically. Local catastrophes might explain a few small graveyards, but certainly not mountains of graveyards and certainly not the abundance of them

worldwide. These were created by the tectonic uplift and geologic energy of the Genesis Flood.

2. There is a general order in the fossil record with lots of exceptions. The sea creatures tend to be buried on the bottom. This would make sense as the breaking up of the fountains of the great deep was the first part of the flood. Billions of sea creatures would have been buried first. More mobile creatures would follow and finally land creatures would have been overwhelmed last of all.

Quiz #5

1. The Geologic Time Table is a man-made hypothetical arrangement of rock layers over vast amounts of time units known as Eons, Eras, Periods and Epochs. The structure is misleading because it does not exist anywhere on earth.

2. The Geologic Time Table was developed over several years. Originally rock formations and layers were studied locally. As questions started to be asked as to their meaning and significance, instead of looking to the Book of Genesis for answers, man formulated an idea that life had progressed over millions of years from simple to complex. The simplest animals were placed on the bottom, the more sophisticated animals developed in later periods of time and belonged on top.

3. An out of place fossil is one that does not belong in a particular layer in which it was found because of the preconception of evolution and deep time.

4. An index fossil is one that is found in a limited time zone (the rock layers). It lived over a particular period of time and then went extinct providing a sort of time marker for its existence and therefore an identification marker for all rock layers which contain that particular fossil.

Quiz #6

1. Answer should include out of place fossils, wrong order, mixed fossils, living fossils, extension of fossil ranges, fossil skipping, and taxonomic manipulations.

2. Index fossils are fossils that are abundant in specific rock layers. Geologists interpret this to mean that they lived for a while in particular geologic time frames, and then went extinct.

Quiz #7

1. Sir Richard Owen, an anatomist and paleontologist who was opposed to Darwin and his ideas.

2. Dragons.

3. The ornithischians meaning *bird-hipped* and the saurischians meaning *lizard-hipped*.

4. Bird and the theropod carnivore, compsagnathus.

Quiz #8

1. Finches.

2. That plants and animals changed within kinds, but not from one kind to another completely different kind.

3. From non-shelled to shelled, invertebrate to vertebrate, fish to amphibian, amphibian to reptile, reptile to mammal and mammal to man.

Quiz #9

Man, male and female, was created in the image of God; different from all other plants and animals. Man could communicate with God. Man could make choices. Man could appreciate good and evil.

Quiz #10

1. Imaginary drawings.

2. Any one of the stories discussed in Lesson 10.

Quiz #11

1. The outline should include a brief account of creation, the Fall, the Flood, the Tower of Babel incident, the post flood ice age, how cave men fit into this picture, the dispersion of man, the found of Egypt and Assyria, the calling of Abraham, the promise of land to Abraham's descendants, the promise of a Messiah from Abraham's descendants, brief history of Israel the Law and sacrifices, the coming and rejection of the Messiah, the ultimate sacrifice of the Lamb, the rise of anti-Christ, the Second Coming of Christ, Christ's kingdom on earth, the last rise of the serpent, the wiping away of the present space and Earth, the ushering of eternity and a new heaven and earth

2. Champollion misidentified Sheshonk I as Shishak of 2 Chronicles 12.

3. Answers may vary somewhat on this, but essentially, the implication would be that the Bible cannot be trusted as to matters of history and dating. The Pyramids, for example, would be dated to have been built before the Flood, yet they were not destroyed by it. Also, the Biblical genealogies are crucial to the establishment of a Messiah, and these would now not be trustworthy, because the dating would be off.

Appendix E – Answers to Comprehensive Exam

1. Champollion
2. Lyell
3. Deism
4. Inventor of dinosaurian
5. Promoted Archaeopteryx as missing link
6. Earth history is cyclical
7. Saurischian
8. Ornithischian
9. Java Man
10. Nebraska Man
11. Piltdown Man
12. Geologic Time Table
13. Fossil
14. Gaps between kinds
15. Cro-Magnon Man
16. Rhodesia Man
17. Peking Man
18. *out of nothing*
19. What Darwin studied
20. Haeckel's drawings
21. Theropod
22. 6th Day of Creation
23. Beginning of the Genesis Flood
24. Tower of Babel incident
25. A post-flood event
26. Considered the "gold standard" of chronologies
27. Masses of fossil bones or shells buried together
28. First multicellular life
29. Bottom rock layers with bizarre looking fossils
30. Sea Lily

Credits

3.0, 32; Cambrian: Ocean scene: http://cnx.org/contents/0a0b301a-b084-4d30-82d5-430c4d266f35@5/The_Evolutionary_History_of_th, http://creativecommons.org/licenses/by-sa/4.0/legalcode, CC by-SA 4.0, 32; Ediacaren fossil: Photo by Verisimilus, http://en.wikipedia.org/wiki/File:Ediacaran_trace_fossil.jpg, http://commons.wikimedia.org/wiki/Commons:GNU_Free_Documentation_License, 33; Ediacaren fossil: Photo by Smith609, http://cnx.org/contents/0a0b301a-b084-4d30-82d5-430c4d266f35@5/The_Evolutionary_History_of_th, http://creativecommons.org/licenses/by/3.0/legalcode, CC by-3.0, 33; Yorgia wagoneri trace fossil: Photo by Aleksey Nagovitsyn, http://commons.wikimedia.org/wiki/File:Yorgia_trace.jpg, http://creativecommons.org/licenses/by-sa/3.0/legalcode, CC by-SA 3.0, 33; Sea Pen: Photo by Nick Hobgood. http://es.wikipedia.org/wiki/Virgulariidae, http://es.wikipedia.org/wiki/Wikipedia:Texto_de_la_Licencia_Creative_Commons_Atribuci%C3%B3n-CompartirIgual_3.0_Unported, CC by-SA 3.0, 34; Sea pen fossil. Photo by Ryan Somma. http://www.flickr.com/photos/ideonexus/2238195678/, https://creativecommons.org/licenses/by-sa/2.0/legalcode, CC by-SA 2.0, 34; Insect: Photo courtesy of Tim and Candey. earths.ancient.gifts, 35; Fish; Photo courtesy of Tim and Candey. earths.ancient.gifts, 35; Trilobite eyes; Photo courtesy of Oscar Sanchez. Bolivian Fossils, 35.

Lesson 3

Painting: By Ivan Ayvazovsky, public domain, http://19thcenturyrusspaint.blogspot.co.uk/2012/09/ivan-aivazovsky-ctd.html, 37.

Lesson 4

Sea: Photo by Walter Baxter, http://www.geograph.org.uk/photo/3264917, CC by-SA 2.0, 42; We win: Image by Vicki Nurre, 43; Fossil Graveyard: Photo by Vicki Nurre, 44; Whale fossils: Photo by Tom Horton, http://www.flickr.com/photos/further_to_fly/2368661537/ , https://creativecommons.org/licenses/by-sa/2.0/legalcode, CC by-SA 2, 45; Whale fossils: Photo by Tom Horton, https://geolocation.ws/v/W/File:Whale%20fossils%20in%20matrix,%20Wadi%20Al-Hitan.jpg/-/en, https://creativecommons.org/licenses/by-sa/2.0/legalcode, CC by-SA 2, 45; Map: Graphic by Vicki Nurre, 46; Diagram: Thunder Bay Limestone, by Vicki Nurre, 46; Crinoid columnals: "Thundering Burial," by Dr. Andrew Snelling (June 1, 1998.) Answers in Genesis, https://answersingenesis.org/, 46; Columnal crinoids: From "Thundering Burial," by Dr. Andrew Snelling (June 1, 1998.) Answers in Genesis, https://answersingenesis.org/, 46; Image from Varricchio, D.J., Sereno, P.C., Xijin, Z., Lin, T., Wilson, J.A. and Lyon, G.H., Mud-trapped herd captures evidence of distinctive dinosaur sociality, *Acta Palaeontol. Pol.* 53(4):567–578, 2008, ref. 2, 48; Image from Varricchio, D.J., Sereno, P.C., Xijin, Z., Lin, T., Wilson, J.A. and Lyon, G.H., Mud-trapped herd captures evidence of distinctive dinosaur sociality, *Acta Palaeontol. Pol.* 53(4):567–578, 2008, ref. 2, 48; Geological Events of the Genesis Flood: By Vicki Nurre, 50; Ogallala Aquifer: http://earthscienceinmaine.wikispaces.com/13.3+Ground+Water, http://creativecommons.org/licenses/by-sa/3.0/legalcode, CC by-SA 3.0, 51; Great Artesian Basin: Source: National Land and Water Resources Audit 2001, found at http://www.environment.gov.au/node/21889, 52; Monument Valley. Photo by Vicki S. Nurre, 53; Grand Canyon. Photo by B Rosen. http://www.flickr.com/photos/rosengrant/2966470172/, https://creativecommons.org/licenses/by-nd/2.0/legalcode, CC by-ND 2.0, 54; Photo: Andrew Schmidt, http://www.publicdomainpictures.net/view-image.php?image=7516&picture=breaking-wave&large=1, 54.

Lesson 5

Graphic: Vicki Nurre, 56; Genesis flood table: Based on Bliss, Parker, and Gish. 1980. *Fossils: Key to the Present.* Master Books, Colorado Springs, out of print, 57; Geologic Timetable. https://dunbarlabsafety.wikispaces.com/CH+12+Resources, http://creativecommons.org/licenses/by-sa/3.0/legalcode, CC by-SA 3.0, 57; Painting: By Hugues Fourau, public domain, found at http://www.ypsyork.org/resources/yorkshire-scientists-and-innovators/william-smith/, http://creativecommons.org/licenses/by-sa/3.0/legalcode, 59; Engraving: http://www.ypsyork.org/resources/yorkshire-scientists-and-innovators/william-smith/, public domain, 59; Map of Europe: By Vicki S. Nurre, 61; Geologic Column: By Vicki Nurre, 64.

Lesson 6

Charles Lyell: http://scienzedidattica.wikispaces.com/Lyell, CCAS 3.0, 66; Geologic Column: Public domain, from nps.gov, found at http://creationwiki.org/File:Geo_time.JPG, 67; Geologic time table index fossils: Public domain, found at http://en.wikipedia.org/wiki/Index_fossil#mediaviewer/File:Index_fossils.gif, 67. Wollemi pine: Photo by Tamsin Slater, https://www.flickr.com/photos/offchurch-tam/3455676720/, CC by-AS 2.0, https://creativecommons.org/licenses/by-sa/2.0/, 68. Fossil: Photo by Frank Franklin II. http://www.livescience.com/3794-dinosaur-fossil-mammal-stomach.html, 68; Fossils: Photo by Tom Darden and Tom Nappi. http://www.dnr.state.md.us/naturalresource/fall2006/cave.pdf, 69; Petrified log: http://www.flickr.com/photos/22327649@N03/2370553207/, https://creativecommons.org/licenses/by/2.0/legalcode, CC by 2.0, 69; Coelecanthe: http://apenvirotuttle.wikispaces.com/KatieCoelacanth, http://creativecommons.org/licenses/by-sa/3.0/legalcode, CC by-SA 3.0, 70; Coelecanth fossil: http://earthscienceinmaine.wikispaces.com/11.1+Fossils, http://creativecommons.org/licenses/by-sa/3.0/legalcode, CC by-SA 3.0, 70; Velvet worm fossil: http://www.fossilmall.com/Science/Sites/Chengjiang/Onychodictyon-ferox/Onychodictyon.htm, 70; Velvet worm: Photo by Bruno C. Velutini, http://ookaboo.com/o/pictures/picture/21149322/The_velvet_worm_Onychophora_is_closely_r, http://creativecommons.org/licenses/by-sa/2.0/, CC-by-SA 2.0, 70; Alvarezsaurus: Photo, http://en.wikipedia.org/wiki/Haplocheirus, http://en.wikipedia.org/wiki/Wikipedia:Text_of_Creative_Commons_Attribution-ShareAlike_3.0_Unported_License, CC by-SA 3.0, 71; Doliodus problematicus: Photo, http://www.dinosaurier.org/2003/10/02/aeltester-fossiler-hai-gefunden/, 72; Anatolepis. Photo, Young, GC, VN Karatajute-Talimaa& MM Smith (1996), *A possible Late Cambrian vertebrate from Australia.* Nature 383: 810-812. Found at http://palaeos.com/vertebrates/pteraspidomorphi/pteraspidomorphi.html, 72; Tiktaalik: Photo by Matt Mechtley, http://www.flickr.com/photos/mmechtley/5379342787/, https://creativecommons.org/licenses/by-sa/2.0/, 73; Evolution of Tiktaalik. Graphic by

Dave Sousa. http://en.wikipedia.org/wiki/Tiktaalik, http://creativecommons.org/licenses/by-sa/3.0/legalcode, CC by-SA 3.0, 74; Eusthenopteron: Photo, http://en.wikipedia.org/wiki/File:Eusthenopteron_model.jpg, http://creativecommons.org/licenses/by-sa/3.0/legalcode, CC by-SA 3.0, 74; Tiktaalik: Drawing by Zina Deretsky, public domain, found at http://es.wikipedia.org/wiki/Tiktaalik, 74; Ichthyostega: Drawing. http://dinosaurs.wikia.com/wiki/Ichthyostega, http://www.wikia.com/Licensing, CC by-SA 3.0, 74; Eusthenopteron fossil: By Haplochromus. http://en.wikipedia.org/wiki/Eusthenopteron#mediaviewer/File:Eusthenopteron_foordi.jpg, http://creativecommons.org/licenses/by-sa/3.0/legalcode, CC by-SA 3.0, 74; Tiktaalik fossil: Photo by Eduard Sola, http://commons.wikimedia.org/wiki/File:Tiktaalik_Chicago.JPG, http://creativecommons.org/licenses/by-sa/3.0/legalcode, CC by-SA 3.0, 74; Ichthyostega fossil: Photo by Oleg Tarabanov. http://en.wikipedia.org/wiki/Ichthyostega#mediaviewer/File:Skeleton_of_Ichthyostega.JPG, http://creativecommons.org/licenses/by-sa/3.0/legalcode, CC by-SA 3.0, 74; Tracks. Image from *Nature*. http://news.bbc.co.uk/2/hi/science/nature/8443879.stm, 75; Bees nests: Public domain, http://nature.nps.gov/geology/paleontology/pub/grd3_3/pefo1.htm, 76; Amber: Photo by Anders L.Damgaard. http://en.wikipedia.org/wiki/Amber#mediaviewer/File:Amber2.jpg, http://creativecommons.org/licenses/by-sa/3.0/legalcode, CC by-SA 3.0, 77; Photo: Maple leaf. http://www.cafepress.com/+crown_maple_leaf_fossil_art_tile_coaster,399500288, 77; Maple leaf: Photo. http://www.flickr.com/photos/evelynfitzgerald/4171391521/, https://creativecommons.org/licenses/by/2.0/legalcode, CC by 2.0, 77; Fossil salamander: Photo by Sulivan et al. http://www.theguardian.com/science/lost-worlds/2014/mar/04/newly-identified-dinosaur-fauna-sheds-light-on-evolution, 77; Spotted salamander. Photo by Scott Camazine. http://en.wikipedia.org/wiki/Salamander#mediaviewer/File:SpottedSalamander.jpg, http://creativecommons.org/licenses/by-sa/3.0/legalcode, CC by-SA 3.0, 77; Fossil shark; Photo. http://www.fossilmuseum.net/fishfossils/sharkinshark/sharkinshark.htm, 78; Shark: Photo. http://marinebiome101.wikispaces.com/Marine+Biome, http://creativecommons.org/licenses/by-sa/3.0/legalcode, CC by-SA 3.0, 78; Scorpion fly fossil: Image taken in 2010 at Smithsonian National Museum of Natural History - Washington, D.C. Found at http://louisvillefossils.blogspot.com/2011/07/jeholomesopsyche-rasnitsyni-scorpionfly.html, 78; Fossil cricket: Photo, http://www.fossilmuseum.net/Fossil_Sites/GreenRiver/cricket/cricket.htm, 81; Fossil mortality plate: Photo, http://www.thefossilforum.com/index.php/gallery/image/18168-dactylioceras-commune-sowerby-1815mortality-plate-toarcian-south-of-caen-feuguerolles-normandy-france/, 80; Protozoa: Photo, http://coolidgelifesciencetext.wikispaces.com/3a.+The+Six+Kingdoms, http://creativecommons.org/licenses/by-sa/3.0/legalcode, CC by-SA 3.0, 81; Fossil protozoa: Photo by Wilson44691, public domain, found at http://en.wikipedia.org/wiki/Foraminifera#mediaviewer/File:Nummulitids.jpg, 81; Fossil succession: http://www.sciencelearn.org.nz/Contexts/Dating-the-Past/Sci-Media/Images/Using-index-fossils, University of Wikato, 82; Bone beds: Photo by Jonathon S. Garcia, Public domain, http://www.nps.gov/agfo/planyourvisit/indooractivities.htm, 82; Fossil bone bed: http://siriusknotts.wordpress.com/2008/09/08/darwins-dyke-what-the-fossil-record-actually-shows/, 82; Fossil bone bed: http://siriusknotts.wordpress.com/2008/09/08/darwins-dyke-what-the-fossil-record-actually-shows/, 83; Polystrate tree: Photo by Michael C. Rygel, http://en.wikipedia.org/wiki/Polystrate_fossil#mediaviewer/File:Lycopsid_joggins_mcr1.JPG, http://creativecommons.org/licenses/by-sa/3.0/legalcode, CC by-SA 3.0, 83; Image by Vicki Nurre, after Roger Gallup, 83; Graph by Patrick J. Nurre, 84.

Lesson 7

Image: By Vicki S. Nurre, 85: Planetoid crashing into the earth: Painting by Don Davis, public domain, http://fr.wikipedia.org/wiki/Extinction_Cr%C3%A9tac%C3%A9-Tertiaire#mediaviewer/File:Planetoid_crashing_into_primordial_Earth.jpg, 86; Photo of Richard Owen: public domain, found at http://es.wikipedia.org/wiki/Richard_Owen#mediaviewer/File:Richard_Owen.JPG, 87; http://www.flickr.com/photos/hhoyer/5534467622/, https://creativecommons.org/licenses/by-sa/2.0/, CC by SA 2.0, 87; Photo by Christian Haugen, http://www.flickr.com/photos/christianhaugen/3423327200/, https://creativecommons.org/licenses/by/2.0/, CC by 2.0, 87. Acambaro figures: Photo by Fchavez2000, found at http://commons.wikimedia.org/wiki/File:Acambaro080407025.JPG, http://en.wikipedia.org/wiki/GNU_Free_Documentation_License, 88; St. George: Painting by Tilemahos Efthimiadis, http://www.flickr.com/photos/telemax/8384487676/, https://creativecommons.org/licenses/by-sa/2.0/, CC by-SA 2.0, 89; St. George: Painting by Raphael, public domain, found at abcgallery.com, 89;St. George: Painting by Peter Paul Rubens, public domain, found at abcgallery.com, 89; Carolus Linnaeus: Public domain, http://commons.wikimedia.org/wiki/File:LinnaeusWeddingPortrait.jpg, CC by-SA 3.0, 90; Triceratops:http://commons.wikimedia.org/wiki/File:Triceratops_Science_Museum_MN.JPG, CC-by-SA 3.0, 91; Triceratops: Photo by Vicki Nurre, 91; Protoceratops: Photo by Andrew Plumb, http://www.flickr.com/photos/aplumb/6000235657/, https://creativecommons.org/licenses/by-sa/2.0/, CC by-/SA 2.0, 92; Ceratopsian skulls: Photo by skinnylawyer, http://commons.wikimedia.org/wiki/File:Ceratopsian_skulls.jpg, http://creativecommons.org/licenses/by-sa/3.0/, CC by-SA 3.0; Therapod foot: Photo by Vicki Nurre, 94; T Rex: Photo by Vicki Nurre, 94; Sauropod foot: Photo by Vicki Nurre, 94; Lizard foot: Photo by Donna Sutton, http://www.flickr.com/photos/77043400@N00/224131637/, https://creativecommons.org/licenses/by-nd/2.0/, CC by-ND 2.0, 97; Long necks: Photo by Vicki S Nurre, 95; Hadrosaur: Drawn by Frederik Spindler, http://creationwiki.org/File:Corythosaurus.jpghttp://creationwiki.org/CreationWiki:GNU_Free_Documentation_License, 95; Stegosaur: Photo by Vicki S. Nurre, 96; Stegosaur: Photo by Eva K. Found at http://en.wikipedia.org/wiki/Stegosaurus#mediaviewer/File:Stegosaurus_Senckenberg.jpg, http://creativecommons.org/licenses/by-sa/2.5/legalcode, CC by-SA 2.5, 96; Pachycephalosaurus: http://es.wikipedia.org/wiki/Pachycephalosaurus, http://es.wikipedia.org/wiki/Wikipedia:Texto_de_la_Licencia_Creative_Commons_Atribuci%C3%B3n-CompartirIgual_3.0_Unported, CC by-3.0, 97; http://commons.wikimedia.org/wiki/File:Pachycephalosauria_jmallon.jpg, http://creativecommons.org/licenses/by-sa/3.0/, CC by-SA 3.0, 97; Ankylosaurus fossil: Photo by Douggers, http://commons.wikimedia.org/wiki/File:FukuiDinosaurMuseum02.JPG, http://commons.wikimedia.org/wiki/Commons:GNU_Free_Documentation_License,_version_1.2. 97; Ankylosaurus: Photo by Vicki S. Nurre, 97; Armadillo: http://commons.wikimedia.org/wiki/File:Armadillo2.jpg, http://creativecommons.org/licenses/by-sa/3.0/, CC by-SA 3.0, 97; Classification of dinosaurs: Image by Patrick Nurre, 98; Ornithischian hip: http://es.wikipedia.org/wiki/Ornithischia, http://es.wikipedia.org/wiki/Wikipedia:Texto_de_la_Licencia_Creative_Commons_Atribuci%C3%B3n-CompartirIgual_3.0_Unported, 99;

Hadrosaur: Picture by Nobu Tamura, http://commons.wikimedia.org/wiki/File:Velafrons_BW.jpg, http://creativecommons.org/licenses/by/3.0/, CC by-3.0, 99; Stegosaur. Photo by Eva K, found at http://en.wikipedia.org/wiki/Stegosaurus#mediaviewer/File:Stegosaurus_Senckenberg.jpg, http://creativecommons.org/licenses/by-sa/2.5/legalcode, CC by-SA 2.5, 99; Triceratops: Photo by Nielseno, http://commons.wikimedia.org/wiki/File:Triceratops_Science_Museum_MN.JPG, http://commons.wikimedia.org/wiki/Commons:GNU_Free_Documentation_License,_version_1.2CC-by-SA 3.0, 99; Ornithopods: Image by Matt Martyniuk, http://commons.wikimedia.org/wiki/File:Largestornithopods_scale.png, http://commons.wikimedia.org/wiki/Commons:GNU_Free_Documentation_License,_version_1.2, 100; Triceratops: Illustration by Tom Parker, found at http://en.wikipedia.org/wiki/Triceratops#mediaviewer/File:Triceratops_by_Tom_Patker, http://creativecommons.org/licenses/by-sa/3.0/legalcode, CC by-SA 3.0, 100; Ornithopod heads: By Pavel Riha, http://en.wikipedia.org/wiki/Ornithopod#mediaviewer/File:Hadrosauroids.jpg, http://creativecommons.org/licenses/by/3.0/, CC by-3.0, 100. Pachycephalosaur: Drawing by Jordan Mallon, found at http://es.wikipedia.org/wiki/Pachycephalosaurus#mediaviewer/File:Pachycephalosauria_jmallon.jpg, http://creativecommons.org/licenses/by-sa/2.5/legalcode, CC by-SA 2.5, 100; Stegosaur: Photo by Eva K, found at http://en.wikipedia.org/wiki/Stegosaurus#mediaviewer/File:Stegosaurus_Senckenberg.jpg, http://creativecommons.org/licenses/by-sa/2.5/legalcode, CC by-SA 2.5, 100; Ankylosaur: By Mariana Ruiz Vilarreal, public domain, found at http://commons.wikimedia.org/wiki/File:Ankylosaurus_dinosaur.png, 100; Bird evolution: http://cnx.org/contents/284c3c70-af0e-4987-a0cf-ac0e4f729713@3.1:39, http://creativecommons.org/licenses/by/3.0/, CC by 3.0, 101; Bird feet: http://fr.wikipedia.org/wiki/Oiseau, http://creativecommons.org/licenses/by-sa/3.0/deed.fr, CC by-SA 3.0, 101; Ornithopod foot: Photo by Vicki Nurre, 101; Saurischian dinosaurs diagram: Diagram by Patrick Nurre, 102; Saurischian hip: http://es.wikipedia.org/wiki/Saurischia, http://es.wikipedia.org/wiki/Wikipedia:Texto_de_la_Licencia_Creative_Commons_Atribuci%C3%B3n-CompartirIgual_3.0_Unported, CC by-SA 3.0, 103; T-Rex: Photo by Vicki Nurre, 103; Sauropod: Photo by Vicki Nurre, 103; Lizard. Photo by John T. Howard, http://www.flickr.com/photos/johnthoward1961/14464763045/, https://creativecommons.org/licenses/by-sa/2.0/, CC by-SA 2.0, 103; Noah's Ark toy: Photo by Tom, http://www.flickr.com/photos/tom1231/3563339788 /, https://creativecommons.org/licenses/by/2.0/, CC by-2.0, 105; Titanic, Noah's ark: Graphic by Vicki Nurre, 105; Hadrosaur egg: Photo by Vicki Nurre, 106; Hadrosaur: By Debivort, http://es.wikipedia.org/wiki/Hadrosauridae#mediaviewer/File:Hypacrosaurus-v2.jpg, http://creativecommons.org/licenses/by-sa/3.0/deed.fr, CC by-SA 3.0, 106; Baby crocodile: Photo by Tim Donovan, http://www.flickr.com/photos/myfwcmedia/6962434859/, https://creativecommons.org/licenses/by-nd/2.0/CC by-ND 2.0, 106; Crocodile: http://commons.wikimedia.org/wiki/File:Crocodylus_acutus_mexico_02-edit1.jpg, http://creativecommons.org/licenses/by-sa/3.0/, CC by SA 3.0, 106;Diplodocus: https://es.wikipedia.org/wiki/Diplodocus, https://es.wikipedia.org/wiki/Wikipedia:Texto_de_la_Licencia_Creative_Commons_Atribuci%C3%B3n-CompartirIgual_3.0_Unported, CC by-SA 3.0, 107; Mososaur: http://es.wikipedia.org/wiki/Tylosaurus, http://es.wikipedia.org/wiki/Wikipedia:Texto_de_la_Licencia_Creative_Commons_Atribuci%C3%B3n-CompartirIgual_3.0_Unported, CC by-SA 3.0, 108; Megalodon jaw: Public domain, found at http://en.wikipedia.org/wiki/Megalodon#mediaviewer/File:Carcharodon_megalodon.jpg, http://es.wikipedia.org/wiki/Wikipedia:Texto_de_la_Licencia_Creative_Commons_Atribuci%C3%B3n-CompartirIgual_3.0_Unported, CC by-SA 3.0, 108; Megalodon comparison: http://en.wikipedia.org/wiki/Megalodon#mediaviewer/File:Megalodon_scale.svg, http://en.wikipedia.org/wiki/Wikipedia:Text_of_Creative_Commons_Attribution-ShareAlike_3.0_Unported_License, CC by-SA 3.0, 108; Megalodon tooth: Photo, http://en.wikipedia.org/wiki/Megalodon#mediaviewer/File:Megalodon_tooth_with_great_white_sharks_teeth-3-2.jpg, http://creativecommons.org/licenses/by-sa/3.0/legalcode, CC by-SA 3.0, 108; St. George and the dragon: Public domain, found at http://fr.wikipedia.org/wiki/Saint_Georges_et_le_Dragon_(Rapha%C3%ABl,_mus%C3%A9e_du_Louvre), http://es.wikipedia.org/wiki/Wikipedia:Texto_de_la_Licencia_Creative_Commons_Atribuci%C3%B3n-CompartirIgual_3.0_Unported, CC by-SA 3.0, 109; St. George and the Dragon: Public domain, found at http://fr.wikipedia.org/wiki/Saint_Georges_et_le_Dragon_(Rapha%C3%ABl,_mus%C3%A9e_du_Louvre), http://es.wikipedia.org/wiki/Wikipedia:Texto_de_la_Licencia_Creative_Commons_Atribuci%C3%B3n-CompartirIgual_3.0_Unported, CC by-SA 3.0, 109; Woolly mammoth: http://fr.wikipedia.org/wiki/Mammouth_laineux, http://creativecommons.org/licenses/by-sa/3.0/deed.fr, CC by-SA 3.0, 110; Thomas Huxley: http://es.wikipedia.org/wiki/Thomas_Henry_Huxley, http://es.wikipedia.org/wiki/Wikipedia:Texto_de_la_Licencia_Creative_Commons_Atribuci%C3%B3n-CompartirIgual_3.0_Unported, CC by-SA 3.0, 110; Compsagnathus skeleton: Photo by Ballista, http://commons.wikimedia.org/wiki/File:Compsognathus_longipes_cast2.jpg, http://creativecommons.org/licenses/by-sa/3.0/deed.fr, CC by-SA 3.0, 110; Archaeopteryx: http://commons.wikimedia.org/wiki/File:Archaeopteryx2_NT.jpg, http://creativecommons.org/licenses/by-sa/3.0/, CC by-SA 3.0, 110; Evolution of flight: Kurochkin, E. N., and I. A. Bogdanovich. 2008, *On the origin of avian flight: compromise and system approaches*, Biology Bulletin 35: 1-11, found at http://people.eku.edu/ritchisong/feather_evolution.htm, 111; Bird evolution: http://zburian.blogspot.com/, 111. Bambiraptor fossil: http://en.wikipedia.org/wiki/Bambiraptor, http://en.wikipedia.org/wiki/Wikipedia:Text_of_Creative_Commons_Attribution-ShareAlike_3.0_Unported_License, CC by-SA 3.0, 112; Bambiraptor fossil: http://es.wikipedia.org/wiki/Bambiraptor, http://en.wikipedia.org/wiki/Wikipedia:Text_of_Creative_Commons_Attribution-ShareAlike_3.0_Unported_License, CC by-SA 3.0, 112; Bambiraptor reconsgtruction: Image by diveofficer, http://www.flickr.com/photos/diveofficer/2913340766/, https://creativecommons.org/licenses/by/2.0/, CC by 2.0, 112; Deinonychus: https://es.wikipedia.org/wiki/Velociraptor, https://es.wikipedia.org/wiki/Wikipedia:Texto_de_la_Licencia_Creative_Commons_Atribuci%C3%B3n-CompartirIgual_3.0_Unported, CC by-SA 3.0, 113; Feathered Deinonychus: Photo by Vicki S. Nurre, 113; Archaeopteryx: Photo by H. Raab, http://en.wikipedia.org/wiki/Archaeopteryx, http://en.wikipedia.org/wiki/Wikipedia:Text_of_Creative_Commons_Attribution-ShareAlike_3.0_Unported_License, CC by-SA 3.0 Unported License, 113; Bird evolution: From a picture by National Geographic Society, found at http://www.pbs.org/wgbh/evolution/library/03/4/image_pop/l_034_01.html, 114. Bird evolution with dates: After a picture by National

Geographic Society, found at http://www.pbs.org/wgbh/evolution/library/03/4/image_pop/l_034_01.html, 115; Biblical geologic column: By Patrick Nurre, 116.

Lesson 8

Finch beaks: Public domain, http://en.wikipedia.org/wiki/Darwin%27s_finches, http://en.wikipedia.org/wiki/Wikipedia:Text_of_Creative_Commons_Attribution-ShareAlike_3.0_Unported_License, CC by-SA 3.0 Unported License, 118; Archaeopteryx: Photo by H. Raab, http://en.wikipedia.org/wiki/Archaeopteryx, http://en.wikipedia.org/wiki/Wikipedia:Text_of_Creative_Commons_Attribution-ShareAlike_3.0_Unported_License, CC by-SA 3.0 Unported License, 121; Archaeopteryx: By Nobu Tamura, http://commons.wikimedia.org/wiki/File:Archaeopteryx_NT.jpg, http://creativecommons.org/licenses/by-sa/3.0/legalcode, CC by-SA 3,0, 121; Archaeopteryx: By EWillougby, found at http://www.mewarnaigambar.web.id/2013/04/mewarnai-gambar-dinosaurus-archaeopteryx.html, 121; Archaeopteryx: Drawing: http://www.itsnature.org/rip/dinosaurs/archaeopteryx/, 121; Archaeopteryx: Drawing: http://www.rmrp.info/archaeopteryx.htm, 121; Picture by NoboTamura, http://paleoexhibit.blogspot.com/, http://creativecommons.org/licenses/by-sa/3.0/, CC by-SA 3.0 Unported, 122. Whale: Public domain, http://en.wikipedia.org/wiki/List_of_cetaceans#mediaviewer/File:Eubalaena_japonica_drawing.jpg, 122; Whale: Public domain, http://en.wikipedia.org/wiki/List_of_cetaceans#mediaviewer/File:Bluewhale877.jpg, 122; Whale: Public domain, http://en.wikipedia.org/wiki/List_of_cetaceans#mediaviewer/File:Humpback_Whale_underwater_shot.jpg, 122; Darwin Day: http://www.uww.edu/news/archive/2012-02-darwin, 126; Whale Evolution: http://coast.noaa.gov/psc/seamedia/Presentations/PDFs/Grade%204%20Unit%204%20Less;on%201%20Whale%20Evolution.pdf?redirect=30locm, 122; Fossil jellyfish: Photo credit University of Kansas, http://www.nsf.gov/news/news_images.jsp?cntn_id=110511&org=NSF, 125; Trilobite: http://commons.wikimedia.org/wiki/File:Naturhistorisches_Museum_Vienna_Dez_2006_071.jpg, http://creativecommons.org/licenses/by-sa/3.0/, CC by-SA 3.0, unported, 125; Trilobite: Photo by Vicki Nurre, 125; Trilobite: Photo by Vicki Nurre, 125; Fish fossil. Photo by Tim and Candey. earths.ancient.gifts, 126; Fossil amphibian: Photo, http://es.wikipedia.org/wiki/Karaurus, http://es.wikipedia.org/wiki/Wikipedia:Texto_de_la_Licencia_Creative_Commons_Atribuci%C3%B3n-CompartirIgual_3.0_Unported, CC by-SA 3.0, 126; Photo: Fossil snake, http://smithlhhsb122.wikispaces.com/Emiley+M., http://creativecommons.org/licenses/by-sa/3.0/legalcode, CC by-SA 3.0, 126; Fish fin: Drawing, http://www.geol.umd.edu/~jmerck/honr219d/notes/09d.html, 126; Frog foot: Drawing. http://www.thunderboltkids.co.za/Grade5/01-life-and-living/chapter2.html, http://creativecommons.org/licenses/by-nd/3.0/legalcode, CC by-ND 3.0, 126; Dinosaur fossil: Photo by rickpilot_2000, http://www.flickr.com/photos/26531284@N02/7868841728/in/photostream/, https://creativecommons.org/licenses/by/2.0/, CC by 2.0, 127; Hyrachus: Photo, https://es.wikipedia.org/wiki/Hyrachyus, https://es.wikipedia.org/wiki/Wikipedia:Texto_de_la_Licencia_Creative_Commons_Atribuci%C3%B3n-CompartirIgual_3.0_Unported, CC by-SA 3.0 Unported, 127; Nebraska Man: Public domain, illustration by Amedee Forestier, http://en.wikipedia.org/wiki/Nebraska_Man, http://en.wikipedia.org/wiki/Wikipedia:Text_of_Creative_Commons_Attribution-ShareAlike_3.0_Unported_License, CC by-SA 3.0 Unported, 128; Teeth: Public domain, http://en.wikipedia.org/wiki/Nebraska_Man#mediaviewer/File:Nebraska_Man_Tooth.jpg, http://en.wikipedia.org/wiki/Wikipedia: Text_of_Creative_Commons_Attribution-ShareAlike_3.0_Unported_License, CC by-SA 3.0 Unported, 128; Horse series: Drawing, http://cnx.org/contents/459d5dd4-cd1e-485d-8006-17a79bdc45b1@2/Evidence_of_Evolution, http://creativecommons.org/licenses/by/4.0/legalcode, CC by-SA 4.0, 129; Horse series: http://immelendoftheyear2012.wikispaces.com/2nd+Period+Unit+5+Evolution+Brooke+and+Hansen, http://creativecommons.org/licenses/by/4.0/legalcode, CC by-SA 4.0, 130; Horse series: http://en.wikipedia.org/wiki/Evolution_of_the_horse#mediaviewer/File:Horseevolution.png Text_of_Creative_Commons_Attribution-ShareAlike_3.0_Unported_License, CC by-SA 3.0, 130; Dinosaur Genealogy: Image by Vicki Nurre, 133.

Lesson 9

Painting: "Storming of Tuileries," by Henri-Paul Motte, public domain, http://fr.wikipedia.org/wiki/Karl_Josef_von_Bachmann#mediaviewer/File:Tuileries_Henri_Motte.jpg, 134;Guillotine: Public domain, found at http://fr.wikipedia.org/wiki/Guillotine#mediaviewer/File:Guillotine_JB_Louvion.jpg, 134; Isaac Newton: Painting by Sir Godfrey Kneller, public domain, found at http://upload.wikimedia.org/wikipedia/commons/3/39/GodfreyKneller-IsaacNewton-1689.jpg, 136; Johannes Kepler: Painting by unknown artist, public domain, http://upload.wikimedia.org/wikipedia/commons/d/d4/Johannes_Kepler_1610.jpg, 136; Galileo: Painting by Justus Sustermans, public domain, http://upload.wikimedia.org/wikipedia/commons/d/d4/Justus_Sustermans_-_Portrait_of_Galileo_Galilei%2C_1636.jpg, 136; Adam and Eve: Painting by Gustav Dore, public domain, found at http://fr.wikipedia.org/wiki/R%C3%A9sum%C3%A9_de_la_Gen%C3%A8se, 137.

Lesson 10

Ernst Haekel: Photo, public domain, http://upload.wikimedia.org/wikipedia/commons/2/2f/Ernst_Haeckel_2.jpg, CC by-SA 3.0, 140; Family tree: Ernst Haekel, public domain, found at http://fr.wikipedia.org/wiki/Ernst_Haeckel#mediaviewer/File:Pedigree_of_man_(Haeckel_1874).jpg, CC by-SA 3.0, 142; Skeleton comparisons: By Waterhouse Hawkins, public domain, found at http://es.wikipedia.org/wiki/Origen_del_hombre#mediaviewer/File:Huxley_-_Mans_Place_in_Nature.jpg, CC by-SA 3.0 unported, 143; Thomas Huxley: Photo, public domain, found at http://es.wikipedia.org/wiki/Thomas_Henry_Huxley#mediaviewer/File:T.H.Huxley(Woodburytype).jpg, CC by SA 3.0 unported, 144; Evolution of man: Drawing, *World Book Encyclopedia 1996*, found at http://www.purifiedbyfaith.com/CreationEvolution/Genesis2/Gen2%20-%20Did%20Man%20Descend%20from%20Adam%20or%20Apes.htm, 144; Neanderthal Man: Hermann Schaaffhaussen, public domain, http://es.wikipedia.org/wiki/Feldhofer, CC by-SA 3.0 unported, 145;Neanderthal Man: Reconstruction by John Gurche, photo by Tim Evanson, http://www.flickr.com/photos/23165290@N00/7283199754/, https://creativecommons.org/licenses/by-sa/2.0/, CC by-SA 2.0, 145; Neanderthal Man: Photo by Photaro, http://en.wikipedia.org/wiki/Neanderthal#mediaviewer/File:Skeleton_and_restoration_model_of_Neanderthal_La_Ferrassie_1.jpg, CC by-SA 3.0, 145; Neanderthal Man: Photo by Stefan Scheer, http://commons.wikimedia.org/wiki/File:Neandertaler_reconst.jpg, http://creativecommons.org/licenses/by-sa/3.0/, CC by SA 3.0, 145; Neanderthal Man: Photo by Jacob Enos, http://farm3.staticflickr.com/2294/2394166246_ac8e2bf29a_z.jpg, https://creativecommons.org/licenses/by/2.0/, CC by-2.0. 145; Calaveras Skull: Originally published in *Review of the Evidence Relating to Auriferous Gravel Man in California* in the Smithsonian Report for 1899, pages 419–472, Plates I-XVI. Washington: Government Printing Office, 1901, public domain, found at http://en.wikipedia.org/wiki/Calaveras_Skull#mediaviewer/File:Calaveras_Skull.jpg, CC by-SA 3.0 unported, 146; Cave painting: found at http://commons.wikimedia.org/wiki/File:Lascaux_painting.jpg, http://creativecommons.org/licenses/by-sa/3.0/, CC by-SA 3.0, 147; Cro-Magnon skull: Public domain, found at http://es.wikipedia.org/wiki/Hombre_de_Cro-Magnon#mediaviewer/File:Cro-Magnon-female_Skull.png, 147; Java Man: Photo, public domain, http://www.talkorigins.org/faqs/homs/wadjak.html, 148; Fossil bones: Photo, public domain, http://en.wikipedia.org/wiki/Java_Man#mediaviewer/File:Pithecanthropus-erectus.jpg, http://creativecommons.org/licenses/by/3.0/, CC by-SA 3.0 unported, 148; Java Man: Photo, public domain, reconstruction 1922, http://es.wikipedia.org/wiki/Homo_erectus_erectus, http://creativecommons.org/licenses/by-sa/3.0/, CC by-SA 3.0 unported, 148; Java Man: Drawings, public domain, found at http://www.harunyahya.com/en/Books/974/the-evolution-deceit/chapter/3573, 148; Java Man drawings: Found at http://www.harunyahya.com/en/Books/974/the-evolution-deceit/chapter/3573, 148; Piltdown Man: Photo by Mike Peel, www.mikepeel.net, http://en.wikipedia.org/wiki/Piltdown_Man#mediaviewer/File:Sterkfontein_Caves_1.jpg, http://creativecommons.org/licenses/by-sa/4.0/, CC-by SA 4.0, 149; Piltdown man: Drawing, public domain, by Charles Henry Bourne Quennell, http://www.conservapedia.com/File:Piltdownman_quennell_1922.png, 149; Piltdown Man: Drawing, public domain, published New York Times, 1921, http://publicdomainclip-art.blogspot.com/2010/12/piltdown-man.html, 149; Piltdown Man representation with orangutan jaw: http://yearsevensealsose.wikispaces.com/Fluorine+Dating, http://creativecommons.org/licenses/by-sa/3.0/, CC by-SA 3.0, 150; Rhodesia Man: Public domain, http://en.wikipedia.org/wiki/File:Rhodesian_Man.jpg, 151; Rhodesia Man: Drawing by Amedee Forestier, public domain, http://en.wikipedia.org/wiki/File:Rhodesian_Men.jpg, 151; Teeth: Public domain, http://commons.wikimedia.org/wiki/File:Nebraska_Man_Tooth.jpg, http://creativecommons.org/licenses/by-sa/3.0/legalcode, CC by SA 3.0 unported, 152; Nebraska Man: By Amedee Forestier,, public domain, found at http://en.wikipedia.org/wiki/Nebraska_Man#mediaviewer/File:Forestier_Nebraska_Man_1922.jpg, 152; Peking Man bones: http://www.ingridpitt.net/archaeology/peking-man.html, 153; Peking Man cast: Photo, http://diogenes.hubpages.com/hub/The-Fossil-Recordand-You, 153; Peking Man: Photo by Mutt, http://en.wikipedia.org/wiki/Peking_Man#mediaviewer/File:Peking_Man.jpg, http://creativecommons.org/licenses/by-sa/3.0/legalcode, CC by SA 3.0, 153; Peking Man: Model by Cicero Moraes, http://en.wikipedia.org/wiki/Peking_Man#mediaviewer/File:Homo_erectus_pekinensis_-_archeaeological.png, http://creativecommons.org/licenses/by-sa/3.0/legalcode, CC by-SA 3.0, 153; Lucy: http://en.wikipedia.org/wiki/Lucy_(Australopithecus)#mediaviewer/File:Lucy_blackbg.jpg, http://creativecommons.org/licenses/by/2.5/legalcodeCC by 2.5, 154; Tools: http://en.wikipedia.org/wiki/Stone_tool, http://creativecommons.org/licenses/by-sa/2.5/legalcode, CC by SA 2.5, 156; Tools: http://en.wikipedia.org/wiki/File:Yiftahel_Pre-Pottery_Neolithic_B_flint_arrowheads.jpg, http://creativecommons.org/licenses/by-sa/3.0/legalcode, CC by SA 3.0, 156; Cave art: Public domain, http://en.wikipedia.org/wiki/Neolithic, http://creativecommons.org/licenses/by-sa/3.0/legalcode, CC by SA 3.0, 157; Farm equipment: Photo by CristianChirita, http://en.wikipedia.org/wiki/Neolithic, http://creativecommons.org/licenses/by-sa/3.0/legalcode, CC by SA 3.0, 157; Dishes and cookware: Photo by Sandstein, http://en.wikipedia.org/wiki/Neolithic, http://creativecommons.org/licenses/by/3.0/legalcode, CC by SA 3.0, 157.

Lesson 11

Tower of Babel: Public domain, http://commons.wikimedia.org/wiki/File:Tour_de_babel.jpeg, http://creativecommons.org/licenses/by-sa/3.0/legalcode, CC by SA 3.0, 158; Information from Encyclopaedia Britannica: Chart by Vicki Nurre, 160; Noah's Ark floating on the waters of the Deluge: Art, Britannica Online for Kids, Web, 7 Oct. 2014, http://kids.britannica.com/comptons/art-142310, 160; Champollion: Painting by Leon Cogniet, public domain, http://fr.wikipedia.org/wiki/Jean-Fran%c3%a7ois_Champollion, http://creativecommons.org/licenses/by-sa/3.0/legalcode, CC by SA 3.0, 162; Rosetta Stone: Photo, public domain, http://commons.wikimedia.org/wiki/File:Rosetta_Stone_BW.jpeg, http://creativecommons.org/licenses/by-sa/3.0/legalcode, CC by SA 3.0, 162; Timeline: By Vicki S. Nurre, after David Rohl, 163; King Menes: Found at LookLex, http://i-cias.com/e.o/menes.htm. 164; National Bank of Egypt: Public domain, found at http://en.wikipedia.org/wiki/Banque_Misr#mediaviewer/File:Banque_misr.jpg, 164; Stele of King Djet: http://en.wikipedia.org/wiki/Djet, http://creativecommons.org/licenses/by-sa/1.0/legalcode, CC by SA 1.0, 165; Sir Leonard Wooley: Photo, http://es.wikipedia.org/wiki/Leonard_Woolley#mediaviewer/File:Leonard_Woolley_(right)_and_T.E.Lawrence_at_the_British_Museum%27s_Ex cavations_at_Carchemish,_Syria,_in_the_spring_of_1913.jpg, http://es.wikipedia.org/wiki/Wikipedia:Texto_de_la_Licencia_Creative_Commons_Atribuci%C3%B3n-CompartirIgual_3.0_Unported, CC by SA 3.0, 165; Zoser statue: Photo by Jon Bodsworth. http://en.wikipedia.org/wiki/Djoser, http://en.wikipedia.org/wiki/Wikipedia:Text of Creative Commons Attribution-ShareAlike 3.0 Unported License, CC by-SA 3.0, 165; Step Pyramid: http://commons.wikimedia.org/wiki/File:Pyramid_of_Djoser_2010.jpg, http://creativecommons.org/licenses/by-sa/3.0/legalcode, CC by SA 3.0, 165; Painting by Pieter Bruegel the Elder, public domain, http://en.wikipedia.org/wiki/Tower_of_Babel#mediaviewer/File:Pieter_Bruegel_the_Elder_-_The_Tower_of_Babel_(Vienna)_-

_Google_Art_Project_-_edited.jpg, http://en.wikipedia.org/wiki/Wikipedia:Text_of_Creative_Commons_Attribution-ShareAlike_3.0_Unported_License, CC by SA 3.0, 165; Aztec pyramid: Photo, http://aztecinformation.wikispaces.com/The+Aztec+Empire, http://creativecommons.org/licenses/by-sa/3.0/legalcode, CC by-SA 3.0, 165; Khufu: Photo, public domain, http://en.wikipedia.org/wiki/Khufu, http://en.wikipedia.org/wiki/Wikipedia:Text_of_Creative_Commons_Attribution-ShareAlike_3.0_Unported_License, CC by-SA 3.0, 166; Great Pyramid: Photo, http://commons.wikimedia.org/wiki/File:Pyramide_Kheops.JPG, http://creativecommons.org/licenses/by-sa/3.0/legalcode, CC by-SA 3.0, 166; Bent Pyramid: http://commons.wikimedia.org/wiki/File:Snefru's_Bent_Pyramid_in_Dahshur.jpg, http://creativecommons.org/licenses/by-sa/3.0/legalcode, CC by-SA 3.0, 166; Senusret I (Sesostris I): Photo by Keith Schengili-Roberts, http://en.wikipedia.org/wiki/Senusret_I#mediaviewer/File:SesostrisI-AltesMuseum-Berlin.png, http://creativecommons.org/licenses/by-sa/3.0/legalcode, CC by-SA 3.0, 166: Joseph son of Jacob: Painting, http://e100project.wordpress.com/category/old-testament-readings/, 166; Joseph son of Jacob: Photo, http://individual.utoronto.ca/mfkolarcik/jesuit/HelenJacobus.html, 167; Joseph's Canal: Photo by NASA, public domain, http://www.nasa.gov/centers/goddard/news/topstory/2006/volcano_nile.html, 167;

Appendices

Time table: By Vicki S. Nurre, 171; Giant fossil dragonfly. http://beforeitsnews.com/science-and-technology/2013/04/geopicture-of-the-week-giant-dragonfly-fossil-2570684.html, 171; Fossil tree stump, Photo from http://www.nature.nps.gov/geology/parks/flfo/, 172; Photo by Slade Winston, fossil flower, http://commons.wikimedia.org/wiki/File:Eocene_fossil_flower,_Clare_Family_Florissant_Fossil_Quarry,_Florissant,_Colorado,_USA_-_20100807.jpg, http://creativecommons.org/licenses/by-sa/3.0/legalcode, CC by-SA 3.0, 172; Photo of fossil insect, Public domain, http://en.wikipedia.org/wiki/Vespinae#mediaviewer/File:Palaeovespa_florissantia.jpg, http://www.nps.gov/parkoftheweek/photo-120.htm, 172; Coelophysis: Photo by Ryan Somma, found at http://www.flickr.com/photos/ideonexus/2829556930/, http://commons.wikimedia.org/wiki/File:Coelophysis_bauri.jpg, CC by-SA 2.5, 173. Photo of fossil sturgeon by Bruce MacAdam, http://www.flickr.com/photos/bruce_mcadam/6393811951/in/photostream/, https://creativecommons.org/licenses/by-sa/2.0/legalcode, CC by-SA 2.0, 173; Drawing of Lake Sturgeon by Virgil Beck, Courtesy Wisconsin Department of Natural Resources, http://www.flickr.com/photos/widnr/8539787064/, https://creativecommons.org/licenses/by-nd/2.0/legalcode, CC by-ND 2.0, 173; Fossil tuatara. Photo. http://wn.com/sphenodontida/images, 173. Tuatara. http://en.wikipedia.org/wiki/Tuatara#mediaviewer/File:Sphenodon_punctatus_in_Waikanae,_New_Zealand.jpguatara, CC by 2.0, 173. Fossil bird: Found at http://www.scienceworldreport.com/articles/6573/20130501/tiny-winged-fossil-reveals-origins-speedy-swift-hummingbird-flight.htm, May 01, 2013, 174; Hummingbird: Photo by Jez Elliott, found at http://www.flickr.com/photos/polarjez/4285356425/, https://creativecommons.org/licenses/by/2.0/legalcode, CC by 2.0, 174;

Patrick Nurre is from the beautiful state of Montana where as a young boy, he spent many of his Saturdays rockhounding near the Big Horn River. This early interest led him to a lifelong study of the world of geology. His experience has included extensive field study in the Pacific Northwest, the Midwest and Plains states, the Southwestern U.S, and most recently, Israel. Patrick conducts classes and seminars in Seattle, and speaks at numerous homeschool conventions on geology and our young earth. He leads a variety of geology field trips every year, including one to Yellowstone National Park, where he helps families discover the Biblical geology of Yellowstone. Patrick's business, Northwest Treasures, is devoted to producing fine geology specimen kits and curricula from a young earth perspective. Patrick and his wife Vicki have three children and two grandchildren. They live in the Seattle, Washington area.

Northwest Treasures
NorthwestRockAndFossil.com
425-488-6848
northwestexpedition@msn.com

Other books by Patrick Nurre:

Rocks and Minerals for Little Eyes (PreK-3)
Fossils and Dinosaurs for Little Eyes (PreK-3)
Volcanoes for Little Eyes (PreK-3)
Geology for Kids (3-6)
Rock Identification Made Easy (3-12)
Fossil Identification made Easy (3-12)
Mineral identification Made Easy (5-12)
Bedrock Geology (high school)
Rocks and Minerals: The Stuff of the Earth (high school)
Volcanoes, Volcanic Rocks and Earthquakes (high school)
Geology and the Hawaiian Islands
Geology and Our National Parks
The Geology of Yellowstone – A Biblical Guide
Genesis Rock Solid